The Panzerkampfwagen IV, or Pzkpfw IV, was the true workhorse of the German army of the Second World War. Originally developed as a close support vehicle which was intended to accompany infantry units into battle, it very quickly evolved into a tank killer and was continuously up-gunned and up-armoured throughout its service.

Almost 7,500 of the gun tank versions were manufactured and the basic hull and suspension were utilised to produce an array of self-propelled artillery pieces and anti-tank guns. Although not as glamorous as the Panther and Tiger, the Pzkpfw IV was relatively easy and cheap to assemble and the versions which left the factories armed with the 7.5cm L/48 gun were more than a match for the enemy vehicles they encountered on the Western Front in 1944.

This book deals specifically with the tanks that took part in the battles for the Normandy bridgehead and the retreat to the Seine between June and late August 1944 and therefore the emphasis is on the Pzkpfw IV ausf H, the most common type in service in the West at that time. A small number of early models were present and they are mentioned, as is the Pzkpfw IV ausf J which was coming into service at the end of the period covered by this study. Hopefully, this volume will be the first of several to examine this remarkable vehicle and earlier and later models will be covered in future titles in some detail.

An important innovation which occurred during this period and deserves to be mentioned was the introduction of the frei Gliederung concept where transport and support services were stripped from companies and concentrated under battalion or regimental command. This German phrase can be literally translated as 'free organisation' but should not be construed to mean an anything-goes approach.

In addition, new tables of organisation were introduced for Panzer units which affected the numbers of tanks available to a company and these are also examined. As we shall see not all the tank units which operated in Normandy had adopted the changes by the time the fighting began.

This Pzkpfw IV ausf H was assembled by Nibelungenwerk in December 1943 and allocated to Panzer-Regiment 3 of 2.Panzer-Division. Note that this vehicle has the second type of support rail for the hull Schürzen and that the first plate has the cut-out which was later dropped from production. Note also that the plates are numbered which was a common practice in this regiment. Other photographs in this series shown that the company number is 821 and Fahrgestellnummer (Fgst Nr.), or chassis number, is 86984.

NORTH-WESTERN FRANCE, SUMMER 1944

Armoured units are show in the approximate positions they occupied on the evening of 5 June 1944. Stationed in the south of France were 11.Panzer-Division near Bordeaux, 2.SS-Panzer-Division north of Toulouse and 9.Panzer-Division outside Avignon. Oberbefehlshaber West (OB West), Feldmarschall Gerd von Rundstedt, had his headquarters at Fontainebleau, just outside Paris, and was responsible for the defence of western Europe. Yet the tanks of Panzergruppe West, under General Leo Geyr von Schweppenberg, and the divisions of Feldmarschall Erwin Rommels' Heeresgruppe B, could not be moved without Hitler's direct authorisation. Nor did von Rundstedt have any control over Kriegsmarine Gruppe West, including the navy's Marine-Artillerie units, or the aircraft of Luftflotte III. The remnants of 9.SS-Panzer-Division and 10.SS-Panzer-Division, which together made up II.SS-Panzerkorps, were still in the East acting as an operational reserve for Heeresgruppe Nordukraine. Not shown here are the many Infanterie and Fallschirmjäger units which were deployed throughout Normandy and Brittany.

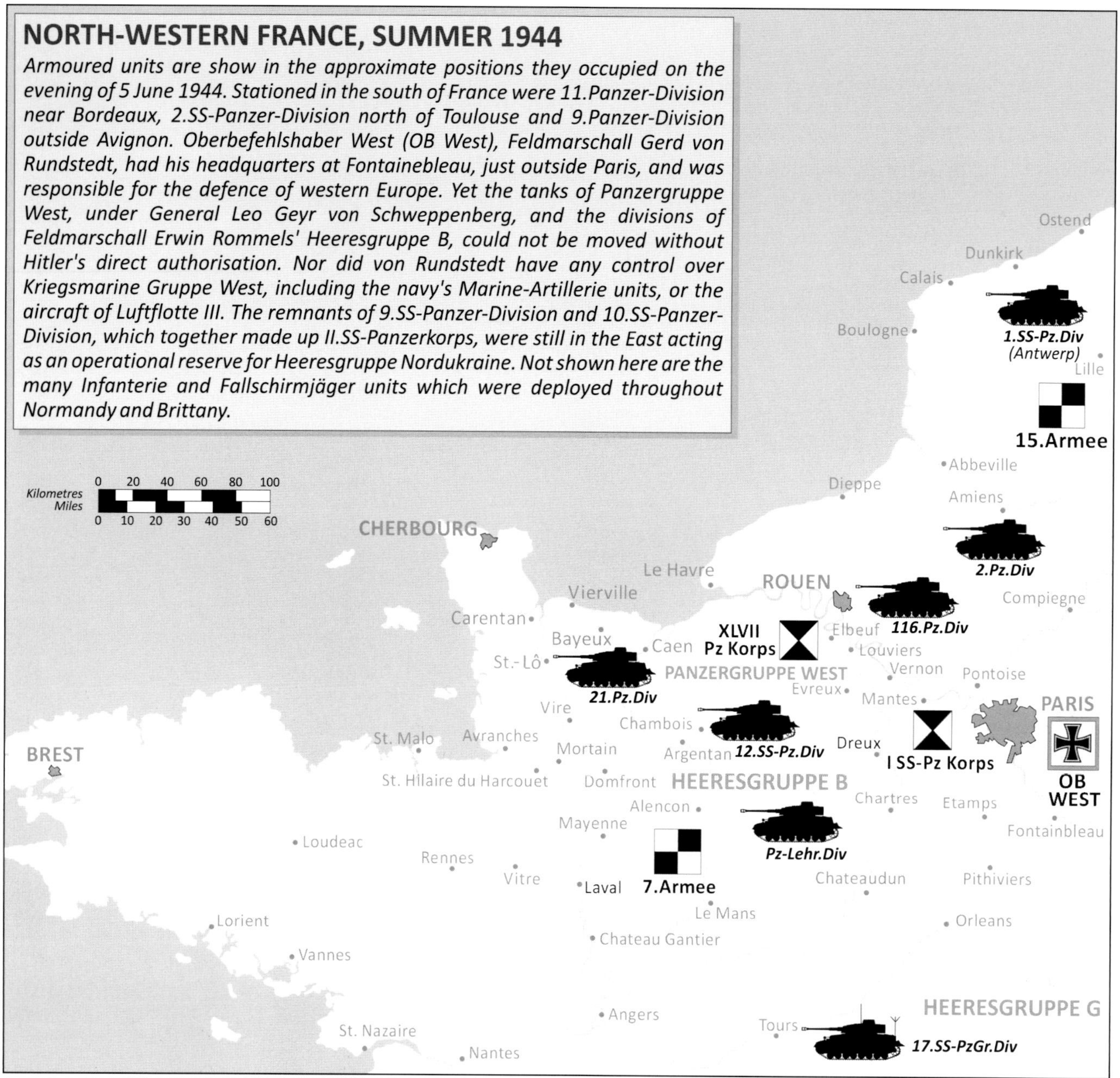

The maps above show the development of the so-called Falaise Pocket between Caen and Argentan, the darker areas indicating the terrain controlled by the Allied armies. Map A depicts the situation at noon on 17 August 1944, the day von Rundstedt's replacement, Feldmarschall von Kluge, was relieved of the command of both Oberbefehlshaber West and Heeresgruppe B. Map B shows the same area with the territory held by the Allies during the evening of 21 August 1944 when the escape corridor through Trun and Chambois was finally closed. The timeline on the following pages gives an outline of the campaign, specifically as it affected the Panzer units.

Photographed in Saint Martin-de-Fresnay, about 20 kilometres south-east of Caen, just prior to the Allied invasion, this Pzkpfw IV ausf B or ausf C was one of six early model tanks on hand with 8.Kompanie, Panzer-Regiment 22 in June 1944. Note the name Hedi painted onto the driver's visor. The buildings in the background have changed little since this image was made.

6 June 1944. *Allied troops land on the Normandy beaches as part of Operation Overlord. Although 21.Panzer-Division, the only armoured formation close to the coast, is officially put on alert at 6.00am, the first tanks would not move off for almost three hours. As the Allied landings are taking place a Kampfgruppe of 12.SS-Panzer-Division is formed and by 10.00am is in action near Buron, some 5 kilometres north-west of Caen. At 5.00pm units of Panzer-Lehr-Regiment 130, stationed near Chartres, receive orders to move towards the front. About 30 minutes later the tanks of Panzer-Regiment 22 begin their counterattack towards Biéville, just 5 kilometres from the coast and as darkness approaches their supporting infantry reach the coast at Lion-sur-Mer, but are almost immediately surrounded. Behind the Panzergrenadiers, the tanks of Panzer-Regiment 22 are held up at Biéville and Périers, reporting the loss of sixteen Pzkpfw IV tanks. Throughout the day the Kampfgruppe from 12.SS-Panzer-Division battles with Canadians units pushing towards Carpiquet airfield, a first-day objective, and manages to hold them up. In the late afternoon the lead elements of Panzer-Lehr-Division are attacked by Allied fighter bombers and lose several vehicles. At midnight the column halts to rest and refuel. Although the defensive effort was at first indecisive and at times ineffective, the town of Caen, another first-day objective, remains in German hands as the day ends.*

7 June 1944. *Between Hérouvillette and Longueval, less than 5 kilometres from the coast, an attack by 4.Kompanie of Panzer-Regiment 22 meets with initial success but is eventually halted by naval gunfire. The Germans are again able to prevent the Canadians from taking Carpiquet airfield, this time pushing the Allies back to the village of Buron, over 5 kilometres from their objective. In a battle fought for the most part in darkness, units of SS-Panzer-Regiment 12 manage to stop a British advance near Franqueville, despite the absence of the regiment's Panther battalion which is stranded on the east bank of the Orne due to a lack of fuel. The tanks of Panzer-Lehr-Regiment 130 continue to struggle towards the front, losing five vehicles to Allied air attacks.*

8 June 1944. *Tanks of SS-Panzer-Regiment 12 assault the Canadian positions near Bretteville-L'Orgueilleuse, a small town on the Caen-Bayeux road. As darkness falls, the tanks of II.Abteilung, Panzer-Lehr-Regiment 130 go into action for the first time at Norrey-en-Bassin and near Brouay. On this day the British and Americans link up north of Bayeux.*

9 June 1944. *By early morning, elements of 2.Panzer-Division, located south of Amiens, begin moving towards the Normandy front. Pounded by offshore naval gunfire the tanks of Panzer-Lehr-Regiment 130 press on towards Bayeux, stopping 5 kilometres south of the city. The commander of II.Abteilung is ordered to prepare his tanks for an assault in support of 21.Panzer-Division and 12.SS-Panzer-Division but they spend the next day trying to seal off a Canadian advance towards Tilly-sur-Seulles. During the day the Luftwaffe makes a rare appearance, first over Lion-sur-Mer and also in an attack on the Orne bridges.*

10 June 1944. *Tanks of Panzer-Regiment 6, attached to Panzer-Lehr-Division, are sent into action for the first time. A single platoon of four tanks manages to halt an enemy advance north of Tilly-sur-Seulles although the British are able to occupy the town that night.*

11 June 1944. *A determined German counterattack drives the British out of Tilly-sur-Seulles but the tanks of Panzer-Lehr-Regiment 130, attacking from Fontenay-le-Pesnel, run into a concentrated artillery barrage and are forced to fall back. Major Prinz Wilhelm von Schönburg-Waldenburg, leading the regiment's II.Abteilung, is killed when his tank is hit by a concealed anti-tank gun. An assault by Canadian tanks near Le Mesnil-Patry, almost exactly halfway between Tilly-sur-Seulles and Caen, is beaten back.*

12 June 1944. *Despite local German successes, British and US Army units are able to link up near Carentan creating a continuous front. Further south, the Americans capture Livry about 20 kilometres east of Saint-Lô.*

13 June 1944. *During the early morning, elements of British 7th Armoured Division reach Villers-Bocage, less than 20 kilometres to the south-west of Caen. At 9.00am they are surprised by a platoon of Tiger I tanks from schwere SS-Panzer-Abteilung 101, later joined by the tanks of Panzer-Lehr-Regiment 130. By the afternoon elements of Panzer-Regiment 3, in their first encounter, had joined the fight and the British withdrew after losing fourteen tanks and two anti-tank guns along with a number of halftracks and lorries. This, and other setbacks suffered on the same day, force the British to temporarily halt their attempts to capture Caen.*

14 June 1944. *British units withdraw from Villiers-Bocage towards Caumont. On the same day US troops complete the capture of Carentan.*

15 June 1944. *Units of Panzer-Regiment 6 attempt to retake the village of Hottot where the men of Panzergrenadier-Lehr-Regiment 902 are surrounded. The attack is called off in the afternoon and the tanks redirected towards Tilly-sur-Seulles.*

16 June 1944. *Kampfgruppe von Luck, made up from elements of 21.Panzer-Division including 4.Kompanie, Panzer-Regiment 22 and Sturmgeschütz-Abteilung 200, attacks St Honorine on the Caen-Carentan road about 5 kilometres south of Arromanches.*

19 June 1944. *Over the next 48 hours the tanks of Panzer-Regiment 3 and Panzer Regiment 6 are involved in heavy fighting around Tilly-sur-Seulles.*

25 June 1944. *The British commence Operation Epsom, an attempt to outflank the German defenders of Caen and secure the city. After two days of costly fighting Caen remains in German hands although the Allies have gained a foothold on the south bank of the Odon River.*

27 June 1944. *The tanks of 4.Kompanie, Panzer-Regiment 22 and SS-Panzer-Regiment 12 mount an attack on Allied positions between Mouen and Cheux, 2 kilometres south of the Carpiquet airfield. Nearby, British armoured units overrun much of Hill 112, the highest point on the Normandy battlefield. Later that day a determined counterattack is made by tanks of SS-Panzer-Regiment 12 and although beaten back, the summit is still controlled by the Germans as night falls.*

29 June 1944. *SS-Panzer-Regiment 9 mounts a major counterattack along the left bank of the Odon River, capturing Grainville-sur-Odon but failing to reach its main objective of Cheux. Anticipating a strong counterattack, all British armoured units around Hill 112 withdraw, allowing the Germans to occupy the area without loss.*

1 July 1944. *Oberbefehlshaber West decides to slowly abandon Caen, seeing the occupation of the city as no longer pivotal to the defence of Normandy. The armoured units around Caen will be gradually transferred to the west to meet the expected American offensive. Of course the British and Canadians know nothing of this and Caen remains the focus of their attention.*

2 July 1944. *Panzer-Lehr-Division is withdrawn from the front line and moved towards the south-east in anticipation of a major American attack.*

9 July 1944. *Tanks of SS-Panzer-Regiment 2 halt an American advance near Périers, north-west of Saint-Lô. British and Canadian units fight their way into Caen and manage to secure the northern suburbs although the Germans stubbornly cling to the southern half.*

A PzKpfw IV ausf H of 5.Kompanie, SS-Panzer-Regiment 12 photographed after the regiment arrived in France but some time before the Allied landings as the right side of this vehicle was badly damaged on 7 June 1944. Note the coating of Zimmerit applied to the hull Schürzen, a practice that may have only been incorporated into production during September and early October 1943. Commanded throughout the Normandy battles by Unterscharführer Willy Kretzschmar, this tank is also shown and discussed on page 24 of the Camouflage & Markings section of this book.

This Pzkpfw IV ausf H of 8.Kompanie, Panzer-Regiment 3 was assembled in January or early February 1944 and other photographs show the Fahrgestellnummer as 89435. This vehicle was built with the hull configuration introduced in January with interlocking armour plates and has Zimmerit applied to the hull sides. Note the notations on the hull Schurzen, a common affectation within this battalion. The three spare track links on the hull rear were a feature of tanks assembled by Nibelungenwerk.

10 July 1944. *After their success of the previous day, the tanks of SS-Panzer-Regiment 2 move forward in a counterattack aimed at the American positions at Saint-Sebastion-de-Raids, two kilometres outside Périers. A major British offensive captures Hill 112 after heavy fighting.*

11 July 1944. *A German counterattack made by units of 9.SS-Panzer-Division drives the British from the summit of Hill 112. To the west, US troops enter Saint-Lô. After fighting almost continuously from the first day of the campaign, 12.SS-Panzer-Division is withdrawn to refit.*

16 July 1944. *A Kampfgruppe made up from elements of Panzer-Lehr-Division and Panzer-Regiment 3 under Oberst Traugott Köhn, the commander of the tank regiment, launches a counterattack to the west of Villiers-Bocage. The Germans are beaten back with heavy loss, including Oberst Köhn who is killed. Tanks of SS-Panzer-Regiment 9 retake the village of Bougy, just off the Caen to Villers-Bocage road.*

18 July 1944. *In a single day of fighting around Bouguébus, south-east of Caen, the British and Canadians lose 100 tanks. Elements of SS-Panzer-Regiment 9, with the infantry of 10.SS-Panzer-Division in support, regain the area between St. Martin, St. André and May-sur-Orne, south of Caen.*

19 July 1944. *Canadian units clear the southern suburbs of Caen. British tanks enter Bras and Hubert-Folie on the outskirts of the town.*

20 July 1944. *The tanks of 1.SS-Panzer-Division launch a major attack west of Bouguébus, parallel to the Caen to Falaise road. On the same day a group of disaffected German officers make an attempt on Hitler's life.*

24 July 1944. *The commander of Panzer-Lehr-Division reports that seven serviceable tanks are available to meet an expected American attack.*

25 July 1944. *The US Army's Operation Cobra begins, preceded by an intense aerial bombardment, and meets with almost immediate success. On the same day the Canadians attack German positions south of Caen. Employing tenacious resistance followed by a series of skilful counterattacks, the Germans manage to stall the Canadian offensive by the evening of the first day. All available tanks of SS-Panzer-Regiment 1 are employed in a counterattack towards Tilly-la-Campagne on the Caen to Falaise road.*

28 July 1944. *American units capture Coutances, about 40 kilometres west of Saint-Lô, but are held up outside the town by elements of 2.SS-Panzer-Division and 17.SS-Panzergrenadier-Division.*

29 July 1944. *A Kampfgruppe made up of two infantry companies and the twenty serviceable tanks of Panzer-Regiment 3 launch an attack towards La Denisière, a small crossroads town on the Villebaudon to Saint-Lô road, in an attempt to link up with Panzer-Lehr-Division and 2.SS-Panzer-Division. But the end of the day, after heavy fighting, the third and fourth companies of SS-Panzer-Regiment 2 are forced to abandon their tanks and flee to the east.*

30 July 1944. *The Panzer-Regiment 3 Kampfgruppe reaches its objective of La Denisière but is quickly surrounded.*

1 August 1944. *Just seven tanks of Panzer-Regiment 3 are able to break out of La Denisière and fight their way back to the German lines at Moyon.*

***2 August 1944**. Tanks of SS-Panzer-Regiment 9 attack British positions at Monchauvet, south-east of Saint-Lô, and although they destroy five British Cromwell tanks, they are cut off and only narrowly escape.*

***4 August 1944.** A Kampfgruppe, made up principally from SS-Panzer-Regiment 9, cuts off a British breakthrough near Chênedollé, destroying thirty-nine enemy tanks. Hill 112 is occupied by British units after the German defenders had abandoned their positions. With the evacuation of Caen the hill has no strategic significance.*

***6 August 1944.** Just before midnight, Operation Lüttich, a major assault intended to blunt the US Army's Cobra offensive and retake Avranches, begins. Despite early tactical gains, including the capture of Mortain, the attack bogs down amidst a determined American counterattack supported by scores of Allied fighter-bombers.*

***8 August 1944.** The Kampfgruppen of 2.SS-Panzer-Division, which had advanced past Mortain towards Saint-Hilaire-du-Harcouët, find themselves in danger of encirclement. On the same day the British begin a major offensive codenamed Totalise which sees fresh Canadian and Polish units attacking towards Falaise in an effort to trap the German units now being pushed eastwards by the Americans. During the late afternoon the Americans occupy Le Mans, the headquarters of SS-Oberst-Gruppenführer Paul Hausser's 7.Armee.*

***10 August 1944.** Near Alençon, north of Le Mans, the tanks of Panzer-Regiment 33 go into action for the first time. Units of 12.SS-Panzer-Division take Point 195, 3 kilometres south-east of Quesnay.*

***11 August 1944.** In an attempt to intercept the German units withdrawing after Operation Lüttich, the British mount an attack towards Mont de Cerisy, east of Vire. Meeting fierce resistance, the attack bogs down outside the village of Le Val after advancing just 800 metres.*

***12 August 1944.** Panzer-Regiment 16 reports that just fifteen tanks are ready for immediate operations.*

***13 August 1944.** The Kampfgruppe made up from the survivors of Panzer-Lehr-Division leave their positions around Habloville, south of Falaise, and withdraw to an assembly area near Fontainebleau. Elements of 9.SS-Panzer-Division reach Vimoutiers, 30 kilometres east of Falaise, after fighting their way through the Allied positions at Trun. The last German units withdraw from Mortain.*

***14 August 1944.** Tanks of SS-Panzer-Regiment 12 halt an Allied advance just north of Falaise but are forced from the defensive positions they have set up by the evening of the next day.*

Photographed in July 1944, this Pzkpfw IV ausf H was assembled by Vomag and has the bolted-on hull armour and side vision ports typical of the early production models. The rails for the hull Schürzen are the first pattern and the Filzbalg oil filters have been lost. Note the early type rear idler, rubber-tyred return rollers and armoured hubcaps on the road wheels. This is one of the tanks handed over to Panzer-Regiment 22 from 3.Kompanie, Panzerjäger-Lehr-Abteilung 130 in late May and is also shown on page 19 of the Camouflage & Markings section.

17 August 1944. *Canadian troops enter Falaise, threatening the German units in Normandy with encirclement. The following morning the Canadians capture Trun, 10 kilometres east of Falaise on the Argentan to Vimoutiers road.*

18 August 1944. *During the early morning the remaining tanks of Panzer-Regiment 3 are withdrawn from the front line and move towards Saint-Lambert, south-east of Villiers-Bocage. That night the last survivors of 12.SS-Panzer-Division escape the pocket which has been formed around Falaise.*

19 August 1944. *Despite a desperate defence, where at one time a single Panther of SS-Panzer-Regiment 2 held up the Allied advance, Polish armoured units link up with US troops at Chambois just south of Mount Ormel as evening approaches. The headquarters of SS-Panzer-Regiment 1, awaiting orders for a counterattack, reports that nineteen tanks are ready for operations.*

20 August 1944. *During the early morning, the remaining tanks of SS-Panzer-Regiment 2 and SS-Panzer-Regiment 9 mount an attack from outside the pocket towards the Polish positions on Mount Ormel. By noon another battle group made up of elements from 10.SS-Panzer-Division, 12.SS-Panzer-Division and 116.Panzer-Division manage to break through the Polish front while SS-Panzer-Regiment 9 prevents Canadian reinforcements from reaching the front. Within a few hours approximately 10,000 Germans are able to escape from the Falaise Pocket. The last tanks of SS-Panzer-Regiment 1 withdraw to the east through the St. Lambert-sur-Dives corridor.*

21 August 1944. *During the early evening Polish and Canadian troops meet at Coudehard, north-east of Trun, closing the German escape route for the last time. Any German units not trapped at Falaise are now in headlong retreat towards the Seine.*

30 August 1944. *The last German troops cross the Seine, leaving much of their heavy equipment behind.*

Said to have been photographed near Saint-Lô, this Pzkpfw IV ausf H of SS-Panzer-Regiment 2 was later towed to Trévières, west of Bayeux, where it was photographed extensively. The turret Schürzen support brackets are of the type used by Krupp and Vomag but the Fahrgestellenummer, which would have given a positive identification, is not evident. Although the company number is only partially visible here the other images show that it was 631 and that the unit insignia of 2.Panzer-Division was carried on the rear hull on the left side. Note the hull Schürzen rails which are of the second pattern with the first triangular hangers trimmed to follow the line of the plates.

Numerically the Pzkpfw IV tank was the most important armoured fighting vehicle in the German army's arsenal and in early June 1944 accounted for 45 to 50 per cent of the front-line tanks available to Oberbefehlshaber West. Almost 600 of these vehicles were stationed in Normandy and the Pas-de-Calais and the tanks of Panzer-Regiment 33, which were fed into the battle in August 1944, with those transferred from Panzer-Regiment 15, would raise that number by more than 100. That total does not include the tanks shipped from the Heereszeugamt (HZA), the army's supply directorate, while the battle was being fought. Presented below are brief histories of the units that operated the Pzkpfw IV tank on the Western Front during the summer of 1944. I have listed the formations of the Heer first, followed by those of the the Waffen-SS, purely as a matter of convenience. The Kriegsstärkenachweisungen (KstN), detailed tables of organisation and equipment, mentioned below are explained in detail in the last section of this book which begins on page 59. The actions of the Panther battalions, discussed briefly here, are covered in more detail in *TankCraft 3: Panther Tanks, German Army and Waffen-SS, Normandy Campaign 1944.*

Panzer-Regiment 3. Attached to 2.Panzer Division, the regiment was withdrawn from the Eastern Front and sent to France during the first weeks of 1944 to be rebuilt. On 20 January a total of forty-five Pzkpfw IV tanks were shipped to the regiment's II.Abteilung followed by a further fifty-three in February. The intention was to outfit the regiment with seventy-nine Panthers and 101 Pzkpfw IV tanks but the full authorisation was never realised and no replacements were received during the fighting in Normandy (1).

On the eve of the Allied landings the division was stationed at Amiens, over 100 kilometres east of Rouen by the most direct route, and reported that ninety-four of the Pzkpfw IV tanks and seventy-three Panthers were available for immediate deployment. Ordered to the front on Friday, 9 June 1944 the first units of 2.Panzer Division were engaged in combat as early as the following Monday but the complete division did not arrive for another week. Importantly these late arrivals included all the division's tanks and most of the self-propelled artillery. Holding the front against British and Canadian units, the division operated in the Caumont area, south-west of Rouen, and on 26 June 1944 the Panthers of I.Abteilung were detached from the regiment to act independently in support of the German defence.

On 21 July the remnants of the division began withdrawing from the front to take part in the subsequent fighting south of Vire during the first week of August 1944, codenamed Operation Lüttich. This was an assault on the US Army positions between Mortain and Sourdeval and the available assets of 2.Panzer-Division were concentrated in a Kampfgruppe led by Oberst Hans Schacke, who had commanded Panzergrenadier-Regiment 304. Made up primarily of the first battalion of Oberst Schacke's regiment, the Kampfgruppe also contained fifteen serviceable Jagdpanzer IV tank destroyers from Panzerjäger-Abteilung 38 and the surviving Pzkpfw IV tanks of Panzer-Regiment 3, although the exact number is uncertain. In addition, the Panthers of Panzer-Regiment 24 had been promised to

Notes

1. This meant that KstN 1175a of 1.11.1943 was used to build II.Abteilung, with twenty-two tanks for each company, while the newer KstN 1177(fG,) dated 1.4.1944, was employed for the I.Abteilung Panthers.

This Pzkpfw ausf H of 8.Kompanie, Panzer-Regiment 3 was disabled and abandoned in the small town of Pont-Farcy on the La Vire River south of Saint-Lô in early August 1944. The unit insignia of 2.Panzer-Division is just visible to the right of the driver's visor and other images of this vehicle show a Fahrgestellenummer 89435. This tank was assembled by Nibelungenwerk in January or February 1944 and was built with the hull with interlocking front plate and side extensions. Note also that the muzzle brake is of the final pattern fitted to these guns. This vehicle is also shown and discussed on page 17 of the Camouflage & Markings section of this book.

Oberst Schacke and elements of 1.SS-Panzer-Division were also under his command. The planned final objective of Avranches was completely unrealistic and by the end of the first day the attack had stalled although the tanks of Panzer-Regiment 3 had pushed up to 10 kilometres into the American defences. By the evening of 7 August 1944 the division had been reduced to almost half its authorised strength.

Retreating towards the east, the last tanks of the division were attached to Kampfgruppe Eberbach and took part in a counterattack near Argentan on Saturday, 12 August 1944. But they were unable to deny the town to the advancing Americans and by the following Friday the survivors were threatened with encirclement in the large pocket which was now developing between Argentan and Falaise.

The division's last significant action took place on 19 August when a Kampfgruppe managed to break through the Canadian front at Lambert-sur-Dives and keep the German escape route open for six hours. None of the tanks survived the Normandy campaign.

Panzer-Regiment 15. Following the French campaign of 1940 the regiment was attached to the newly formed 11.Panzer-Division and, commanded by Oberst Meinrad von Lauchert, was stationed in southern France outside Bergerac, east of Bordeaux, in June 1944. In the same month the division was reinforced with parts of 273.Reserve-Panzer-Division and a number of the fifty-four Pzkpfw IV tanks allocated to that unit were absorbed by II.Abteilung, Panzer-Regiment 15.

On Sunday, 6 August 1944 the division received orders to hand over all the tanks of II.Abteilung to the headquarters of Heeresgruppe B and these were to be used to reinforce the units fighting in Normandy. Major Johann Berger, the battalion commander, was to coordinate the operation, assisted by Oberleutnant Edwin Graf von Francken-Sierstorpff of 8.Kompanie, and a total of sixty-one Pzkpfw IV tanks had been loaded onto railway cars by the following Thursday.

Each tank was accompanied by its commander and driver only and by 12 August the trains had reached Orleans but were forced to continue from there on their own tracks to Chartres and then on to Dreux. There, late in the evening of 15 August 1944, all sixty-one Pzkpfw IV tanks were handed over to a Waffen-SS unit (1).

For some reason the battalion retained sixteen of its Pzkpfw IV tanks and these took part in the fighting in the Rhône Valley, mostly with Oberleutnant Rudolf Graf Hoyos' 6.Kompanie, although as many as four tanks were with the regimental headquarters (2).

Panzer-Regiment 16. This unit was formed in France in late May 1944 and attached to 116.Panzer-Division. When the regiment arrived at the front it had on hand a total of eighty-six Pzkpfw IV tanks and seventy-six Panthers, the latter seconded from Panzer-Regiment 24 (3).

Interestingly a report of 8 June 1944 indicates that three of the Pzkpfw IV vehicles were early models armed with the short 7.5cm gun. The HZA allocation lists show ten Pzkpfw IV tanks shipped to the battalion in April 1944, and then diverted to SS-Panzer-Regiment 1, while a further ten were despatched in May, arriving at the end of the month. The remaining tanks may have come from 179.Reserve-Panzer-Division which was assigned fifty-seven tanks in March and was associated with Reserve-Panzer-Abteilung 1. In any case, the Panzer regiment received no replacement tanks until November 1944 and was forced to fight the battles in Normandy with its ever-dwindling resources. The first elements of the division reached the front south-east of Caen on Monday, 24 July 1944 but on the following Friday the Americans launched Operation Cobra, a huge armoured assault designed to break through the German lines and sweep west, and within days the division was diverted to Vire, over 40 kilometres to the south-west.

In a determined counterattack the tanks of Panzer-Regiment 16 managed to blunt the advance of the US 2nd Armored Division but, hampered by constant air attacks, the Germans were forced to withdraw. In early August the division was attached to XXXXVII.Panzerkorps in preparation for the German counteroffensive at Mortain, the largest tank battle of the Normandy campaign, but the near-treasonous conduct of its commander, General Gerhard Graf von Schwerin, meant that the tanks saw little, or no, action (4).

Retreating towards the east the remnants of 116.Panzer-Division battled to hold open the escape corridor on the southern edge of the Falaise Pocket. But by now only a few tanks remained and when the division's last reserve was committed on 17 August 1944 it consisted of just eighty men from the reconnaissance battalion. During the night of 20 August some elements managed to break out, crossing the Lambert bridge near Coudehard, but many were trapped inside the pocket, their last reported position being between Argentan and Trun on the northern bank of the River Orne. The division's acting commander reported that his formation consisted of a weak infantry battalion and twelve tanks of all types and it is one measure of how desperate the German defence in the west had become that the survivors of Falaise were tasked with holding the Seine crossing at Elbeuf, south of Rouen.

Notes

1. The exact identity of this unit has never been satisfactorily established although a notation in the HZA allocation list that SS-Panzer-Regiment 1 received seventeen Pzkpfw IV tanks from Nachschub West, the army's supply directorate for the Western Front, in early September may provide a partial clue. In his own account, written as recently as 2013, von Francken-Sierstorpff mentioned only 'an SS tank regiment'.

2. These numbers are debated but the total is usually the same. In his *Ihr Glaube galt dem Vaterland: Geschichte des Panzer-Regiments 15,* Gustav Schrodek suggests that these tanks had all undergone significant repair and this may be the reason they were not relinquished in early August.

3. The story of I.Abteilung, the Panther battalion of Panzer-Regiment 16, is rather complex and is explained in detail in *TankCraft 3: Panther Tanks German Army and Waffen-SS, Normandy Campaign 1944.*

4. Graf von Schwerin's insubordination was only matched by his luck. His refusal to order his tanks forward at Mortain led to only a temporary dismissal, ending in late September. He later abandoned the defence of Aachen, after notifying the Americans of his decision to do so, largely through a series of misunderstandings, and he was suspected of involvement in the plot to kill Hitler. But he was merely reprimanded and posted to the Italian Front where he commanded a corps.

Panzer-Regiment 22. From late 1942 and into the first months of 1943 an attempt was made to bring together the units equipped with tanks that had been captured from the French army after the 1940 campaign. This formation was eventually named Panzer-Regiment 100 and was made up of two battalions, each with three companies. By 15 July 1943 the regiment had been attached to the newly-created 21.Panzer-Division and expanded to eight companies and it may have been at about this time that the decision was made to replace the French vehicles with Pzkpfw IV tanks.

The first fifty-three new tanks were shipped from the HZA in September 1943 but it seems that twenty-one of these were diverted to 14.Panzer-Division which was at that time preparing to return to the Eastern Front. The subsequent records are somewhat confusing, with the HZA figures for December stating that the division had fifty-seven Pzkpfw IV tanks on hand although a report given to Hitler at the beginning of the month stated that the planned allocation of seventy-eight of these tanks to 21.Panzer-Division was progressing well with seventy already delivered. I have presumed that the HZA numbers are the more accurate and with a delivery of twenty tanks in January, and a further twenty in February, the total was at that time recorded as ninety-eight.

From 2 May 1944 a series of orders directed that a number of tanks were to be handed over from Panzer-Lehr-Division and although the original directive was amended, by late May nine Pzkpfw IV tanks from Panzer-Regiment 130, four additional vehicles from 3.Kompanie, Panzerjäger-Lehr-Abteilung 130 and a further two tanks from Panzer-Lehr-Nachrichten-Abteilung 130 were with Panzer-Regiment 100. Both the latter were Panzerbefehlswagen IV command tanks. At the same time the Panzer-Regiment was ordered to be reorganised according to the Freie Gliederung concept (1) and renamed Panzer-Regiment 22.

It was planned that I.Abteilung would be made up of four companies each with seventeen Pzkpfw IV tanks while the companies of II.Abteilung would contain fourteen. The French tanks were gradually relinquished, possibly to equip Panzer-Abteilungen 205 and 206, and the crews sent to the training centre at Mailly-le-Camp, east of Paris. Some of these men returned to the division in August 1944 with a number of Pzkpfw IV tanks issued directly from Nachschub West, the only replacements received during the Normandy battles (2).

This regiment was in action from the first day of the invasion and most of its vehicles were lost in the early fighting. On 8 July 1944 the headquarters of 21.Panzer-Division reported that fifty-four Pzkpfw IV tanks had been completely destroyed and within three weeks what remained of II.Abteilung was withdrawn and was not returned to the division until later in the year. Much of the division was trapped inside the Falaise Pocket and by the end of August just twelve tanks remained operational and these fought on for some weeks as a rearguard, supporting the infantrymen of 16.Luftwaffen-Feld-Division.

Panzer-Regiment 33. Attached to 9.Panzer Division, this formation was undergoing training near Nimes in

Notes

1. This is explained further in the chart on page 61.
2. One allocation of seventeen tanks was made on 21 June but not shipped until 8 July 1944. A second consignment of ten vehicles was despatched on 30 July but these are noted as not being with the battalion until 5 September 1944.

Another view of the Pzkpfw IV ausf G of Panzer-Regiment 22 shown in the photograph on the contents page. The first pattern hull Schürzen support rails are clearly visible here.

A Pzkpfw IV ausf H of Panzer-Regiment 33 disabled and abandoned near the town of Sées, 20 kilometres south of Argentan, in late August 1944. The turret Schürzen support brackets confirm that this is a Nibelungenwerk-built tank. Note the rectangular splash guards of the driver's and radio operator's hatches.

southern France when the Allied landings took place and was not ordered to move towards the front until 27 July 1944. At that time the regiment's I.Abteilung, commanded by Hauptmann Ewald Becker, reported that eighty-nine Pzkpfw IV tanks were on hand but II.Abteilung, with its seventy-nine Panthers, was in training and the two battalions did not operate together during the Normandy campaign.

Hampered by constant air attacks and partisan ambushes, it was almost ten days before the tanks could take part in the counterattacks made in an effort to halt the American advance in the first week of August. This regiment was one of the strongest armoured units available to the Germans at the time but was fed piecemeal into the defence of the line between Domfront and Mayenne and the area south of Alençon with I.Abteilung attached to Panzergruppe Eberbach. By 11 August 1944 the battalion had been reduced to twelve tanks and was supported by a few companies of infantry and parts of the division's artillery regiment. By mid-August the division reported that over 100 tanks of all types had been lost and it is likely that the majority of these came from I.Abteilung. Much of 9.Panzer-Division was trapped in the Falaise Pocket but the survivors, including parts of II.Abteilung, were formed into a Kampfgruppe and fought in the defence of the Seine crossings with the remnants of 116.Panzer-Division.

Panzer-Lehr-Regiment 130. This unit was formed in late 1943 in part from personnel of the Panzer-Lehr-Abteilung, a training unit of Panzertruppenschule Wünsdorf.

The regiment's I.Abteilung was formed at the same time and commanded by Major Prinz Wilhelm von Schönburg-Waldenburg, who had led a battalion of Panzer-Regiment 39 until February 1943 when he was badly wounded. In January 1944 the first battalion of Panzer-Lehr-Regiment 130 was renamed II.Abteilung and together with the regimental headquarters was training near Verdun in eastern France. The first shipment of sixteen Pzkpfw IV tanks had been made in late September 1943 and by February of the following year the division reported that a total of ninety-eight had arrived (1). In addition, a few Pzkpfw IV tanks had

Notes

1. As a matter of interest, the Panthers of II.Abteilung did not take part in the Normandy battles and a battalion of Panzer-Regiment 6 was attached to the division. This is dealt with in some detail in *TankCraft 3: Panther Tanks, German Army and Waffen-SS, Normandy Campaign 1944* and *TankCraft 18: Panther Tanks, German Army and Waffen-SS, Defence of the West 1945.*

been brought to France from Bergen where the unit had served as a training formation. No further reinforcements were allocated before the invasion began and those all came directly from Nachschub West, the army's supply directorate in western Europe.

During the first two weeks of May 1944 the regiment was ordered to hand over nine of its Pzkpfw IV tanks to Panzer-Regiment 22 of 21.Panzer-Division and a further six tanks from the signals and Panzerjäger battalions of Panzer-Lehr-Division were also given up (see Panzer-Regiment 22). An order of 2 May instructing Panzer-Artillerie-Regiment 130 to relinquish their Pzkpfw IV tanks was rescinded.

On 1 June 1944, less than a week before the Allied armies landed in Normandy, the regimental headquarters reported that five Pzkpfw IV tanks were on hand while a further eight were with the headquarters of II.Abteilung. Both 5.Kompanie and 6.Kompanie had twenty-one tanks, one less than their authorised allocation, and 7.Kompanie and 8.Kompanie were at full strength with twenty-two.

The division was engaged against the British and Canadian units around Caen until it was gradually withdrawn between 26 June and 5 July 1944 and was replaced in the front line by 276.Infanterie-Division. During that time the Panzer regiment received eleven Pzkpfw IV tanks as replacements. The division returned to the front and on Tuesday, 11 July 1944 took part in the attack against US positions near Le Désert, along the Vire to Villers-Bocage road, and despite some initial successes the German forces were eventually forced to withdraw. Elements of Panzer-Lehr-Division were caught in the massive aerial bombardment that preceded the US Army's Operation Cobra but sources differ over the extent to which the division was affected. Some narratives, including that of Generalleutnant Fritz Bayerlein, the division's commander, go so far as to claim that the division was virtually annihilated as a result of the bombing.

But the new commander of II.Abteilung, Hauptmann Helmut Ritgen (1), not only stated that none of his Pzkpfw IV tanks were lost, as the battalion was held in reserve, but that very few Panthers and Jagdpanzers were destroyed.

On 30 July a total of twelve new Pzkpfw IV tanks were allocated by Nachschub West but these were not shipped until the first week of August and it was 5 September 1944 before they were recorded as being on hand with the regiment. In early August it had been decided to withdraw the bulk of the division, leaving a Kampfgruppe under Oberst Paul Freiherr von Hauser, who had led Panzergrenadier-Lehr-Regiment 901, as a rearguard. Oberst von Hauser had under his command parts of the artillery regiment and the remaining Pzkpfw IV tanks of II.Abteilung, enough to form a weak company-sized unit (2).

On 12 August Kampfgruppe von Hauser was withdrawn and on the following day Bayerlein ordered what remained of his division to follow.

A Kampfgruppe consisting of a company of tanks, a howitzer battery and parts of Panzergrenadier-Lehr-Regiment 902 remained behind and together with remnants of 12.SS-Panzer-Division managed to break out of the Falaise Pocket

Notes

1. Major von Schönburg-Waldenburg had been killed on 11 June 1944.
2. In his post-war account Helmut Ritgen implies that this was 7.Kompanie under Oberleutnant Kues.

This Pzkpfw IV ausf H of 8.Kompanie, Panzer-Lehr-Regiment 130 is one of at least three tanks the regiment lost in the fighting for the town of Saint-Gilles, west of Saint-Lô, in late July 1944. A series of photographs taken at the time show that it was actually disabled in the centre of the town with two tanks of 5.Kompanie and presumably towed to the position seen here. What has been assumed to be additional armour protection on the turret roof is actually one of the hull Schürzen plates. This vehicle is also shown and discussed on page 21 of the Camouflage & Markings section of this book.

on 20 August 1944.Two days later Oberst Rudolf Gerhardt, the regiment's commander, reported that ten Panthers and ten Pzkpfw IV tanks were all that had survived the battles in Normandy.

Panzer-Abteilung 103. Rebuilt in mid-1943 after the original battalion was destroyed at Stalingrad, this formation was attached to 3.Panzergrenadier-Division. In late August 1944 the division was transferred from the Italian Front to France and was involved in the battles along the Marne River and led counterattacks west of Chaumont-Bar-le-Duc and Saint-Mihiel before withdrawing to the area south-west of Metz. There are persistent accounts of this battalion possessing a single Pzkpfw IV tank in August but I have not been able to find any record in the official allocations. It is tempting to speculate that this tank may have been one of the vehicles of 4.Kompanie, Panzer-Abteilung 215 that served with the battalion in Italy.

Panzer-Abteilung 115. Formed in mid-1943 from the first, second and third companies of Panzer-Abteilung 215, the battalion was attached to the recently-formed 15.Panzergrenadier-Division and served in Italy during 1943 and the early part of 1944.

The fourth company was subordinated to Panzer-Abteilung 103 for some time before being disbanded. In August 1944, the division was transferred to France and took part in the battles in Lorraine and later the fighting around Venlo and the defence of Aachen.

On 21 August 1944, shortly after arriving in France, the battalion received thirty-one Pzkpfw IV tanks directly from Nachschub West. It is possible that a few tanks were brought from Italy but I have been unable to confirm this.

Panzer-Abteilung (Funklenk) 302. Formed in July 1944 with three Panzerbefehlswagen IV command tanks, four Sturmgeschütz III assault guns and thirty-six Borgward BIV remotely-controlled demolition vehicles, the battalion served briefly in France and was transferred to the East, taking part in the fighting for Warsaw.

The armoured formations of the Waffen-SS formed an important part of the German defence in the West and for some time bore the brunt of the heaviest fighting of the Normandy campaign. Listed here are the Panzer units of the Waffen-SS that were equipped with the Pzkpfw IV tank.

SS-Panzer-Regiment 1. This unit's parent formation, 1.SS-Panzer-Division Leibstandarte Adolf Hitler (LAH), had been reduced to a weak Kampfgruppe during the winter battles on the Eastern Front and by April 1944 the survivors were withdrawn to Belgium to rest and refit. The division's Panzer regiment was to be organised with two battalions, each made up of four companies. The companies of I.Abteilung would each contain seventeen Panthers while those of II.Abteilung would have twenty-two Pzkpfw IV tanks (1).

In early May 1944, II.Abteilung received thirty-five tanks, a number of which had originally been destined for Panzer-Regiment 22 and Panzer-Regiment 16, and a further fifty-three in mid-June 1944. A notation on the June delivery mentions that ninety-eight Pzkpfw IV tanks were now on hand and from this we may assume that a number of tanks had been brought to the West from Russia (2).

After several false starts the division began moving towards the front on 17 June 1944 but the only major unit to take part in the fighting during that month was SS-Panzergrenadier-Regiment 1 and it was not until the first week of August that the complete Panzer regiment was with the division (3).

The division remained in the area south of Caen fighting alongside 12.SS-Panzer-Division, together forming I.SS-Panzerkorps, until the evening of 5 August 1944 when the remnants of the Panzer regiment, two infantry battalions and other units were detached to support 2.Panzer-Division in Operation Lüttich, the Mortain counterattack. On 22 August, after breaking out of the Falaise Pocket, the division reported no serviceable tanks or artillery pieces were on hand.

The August 1944 allocation list mentions that seventeen Pzkpfw IV tanks were shipped to the regiment directly from Nachschub West but these did not arrive until the first week of September.

SS-Panzer-Regiment 2. Attached to 2.SS-Panzer-Division Das Reich, by early 1944 the regiment had been reduced to a battalion-sized Kampfgruppe made up from the division's surviving Panthers, Tigers and Pzkpfw IV tanks.

In April the remnants of the division were withdrawn to the West and SS-Panzer-Regiment 2 was rebuilt with two tank battalions each of four companies. At that time the division reported that just five Pzkpfw IV tanks were on hand and four of these were in repair. The regiment's I.Abteilung was to be re-equipped with seventeen Panthers in each company while II.Abteilung was to be a Gemischte, or mixed, battalion with two companies of Pzkpfw IV tanks and two companies of Sturmgeschütz III assault guns. On 27 March 1944 a total of fifty Pzkpfw IV tanks were shipped from the HZA and a further twenty-four were sent in June (4).

In the same month the organisation of II.Abteilung was amended to include three companies of Pzkpfw IV tanks and a single company of assault guns.

Notes

1. As explained for Panzer-Regiment 3 (above) this involved a mix of KstN which where current and impending.

2. Twenty Pzkpfw IV tanks had been allocated from Nachschub Ost in January 1944 of which seventeen were shipped.

3. Other important elements of the division where still in Belgium including most of SS-Aufklärungs-Abteilung 1,whose promised Sdkfz 250/9 halftracks never materialised.

4. The HZA allocation list mentions that the June delivery brings the total on hand to seventy-eight and it may be that some tanks were returned from repair.

Notes

1. The authorised allocation was twenty-two vehicles for each company. The figures given here represent the on hand totals and it should be mentioned that several were in need of repair.

2. Interestingly, a report of the same day states that sixty-seven enemy tanks had been captured, eleven of them in running order.

On Sunday, 11 June 1944 the division was placed under the command of Heeresgruppe B and the next day ordered to move to the front. While the division's tanks and other tracked vehicles prepared to move by rail, the wheeled elements, many of which had been commandeered at the last moment, set off by road and by the following Friday had reached the area around Mortain.

Much has been made of the contribution of the French Resistance and Allied air power in delaying the northern movement of the division but in fact the greatest obstacle was the lack of vehicle spare parts. As late as 30 June many units, including a company of Pzkpfw IV tanks of II.Abteilung, were only just leaving southern France for Normandy and it was not until 26 July that the last trains were unloaded.

In early August most of the division's remaining armour assets were formed into two Kampfgruppen, both controlled by Brigadeführer Otto Baum, which took part in Operation Lüttich, capturing the town of Mortain and pushing on to the west and south-west as far as Saint-Hilaire-du-Harcouët, only withdrawing when encirclement was imminent.

But by this time the regiment's combat strength amounted to three Panthers, eight assault guns and five Pzkpfw IV tanks but these still managed to fight units of the 1st Polish Armoured Division to a standstill near Mount Ormel on 20 August 1944, allowing many German units to escape the Falaise Pocket. On 28 August the division was able to report that six serviceable tanks of all types were all that remained.

SS-Panzer-Regiment 9. This unit, and its parent formation 9.SS-Panzer-Division Hohenstaufen, was operating as part of II.SS-Panzerkorps in northern Ukraine when the Allies landed in Normandy. Ordered to the West on 12 June 1944 the regiment's II.Abteilung reported that 5.Kompanie and 6.Kompanie had a total of forty-six Pzkpfw IV tanks on hand while 7.Kompanie and 8.Kompanie were equipped forty Sturmgeschütz III assault guns (1).

The regiment's I.Abteilung was in training at Mailly-le-Camp, east of Paris, and joined the division in the area between Falaise and Condé-sur-Noireau as it moved towards the Normandy front. The regiment went into action for the first time on 29 June 1944 in an attack on Grainville-sur-Odon, 5 kilometres west of Caen, and took the town in the early fighting. But as the tanks pushed on towards Noyer-Bocage they were stopped by a massive air attack and in two days of fighting in this area II.Abteilung lost nine assault guns and eleven tanks completely destroyed. By the first weeks of July the regiment's tanks and assault guns were so badly in need of repair that just twenty vehicles could be employed in a counterattack north of Evrecy and although forty-eight British tanks were knocked out, the regiment lost five of its own. By 18 July the headquarters of 9.SS-Panzer-Division reported that twenty-five Panthers, fifteen assault guns and thirteen Pzkpfw IV tanks were operational (2).

But by the end of July, II.Abteilung reported that twenty-seven assault guns and twenty-two Pzkpfw IV tanks were on

Photographed in late August 1944 in what is today Rue de Dr Prodhomme in the village of Putanges-Pont-Écrepin, south of Falaise, this Pzkpfw IV ausf H may have been one of the tanks of the headquarters of II.Abteilung, SS-Panzer-Regiment 1. Note the Zimmerit on the turret Schürzen, the cast return rollers, later rear idler and mix of hubcap types. Note also the two wooden stowage boxes on the hull side.

At left: A Pzkpfw IV ausf H of 7.Kompanie, SS-Panzer-Regiment 2. Note the field-modified antenna and insulator on the turret side. Another image of this tank shows the three spare track links on the rear hull plate, a feature that was indicative of Nibelungenwerk-built tanks. This Pzkpfw IV is also shown and discussed in the Camouflage & Markings section on page 23. Below: Tanks of 6.Kompanie, SS-Panzer-Regiment 6 photographed during the summer of 1944, probably before the Normandy landings. A piece of film which features these vehicles shows that the tank in the foreground is a Pzkpfw IV ausf J. The hull Schürzen plates and rail are of the second pattern introduced in September 1943 and the cut-out in the first plate was a feature of Schürzen fitted by Nibelungenwerk up to March 1944.

hand and as no allocations were made from the HZA or Nachschub West a number of vehicles must have been returned from repair. The division remained in the area south of Caen until 16 August 1944 when it was ordered to withdraw towards Vimoutier, outside what would become known as the Falaise Pocket. During this move some of the Panzer regiment's tanks were taken over by SS-Panzer-Regiment 12.

The remnants of the Panzer regiment, with elements of 2.SS-Panzer-Division, were involved in the attempts to relieve the encircled German units from the outside. Somewhat surprisingly, the regiment reported that between twenty and twenty-five tanks of all types were on hand at the conclusion of the battles east of Falaise and in early September these were handed over to 11.Panzer-Division.

SS-Panzer-Regiment 10. Attached to 10.SS-Panzer-Division Frundsberg, this unit was also fighting in Russia when the Normandy invasion began. Ordered to move to France on 12 June 1944, the Panzer regiment's II.Abteilung was made up of thirty-eight Sturmgeschütz III assault guns and thirty-nine Pzkpfw IV tanks with three Pzkpfw III command vehicles, the latter almost certainly with the regimental headquarters (1).

It was not until Saturday, 24 June 1944 that the first elements of the division, consisting mainly of the divisional headquarters, the tanks of the Panzer regiment and the reconnaissance battalion, reached the Normandy front. The division first went into action on the following Friday but was restricted by a severe lack of fuel and after four weeks of fighting reported that just twenty assault guns and twenty Pzkpfw IV tanks were available for operations, although a number were expected to return from repair.

During July the division remained in the area south-west of Caen until early August when it was sent to bolster the southern side of the pocket which had formed west of Falaise. Parts of 10.SS-Panzer-Division were trapped in the Falaise Pocket and by 21 August 1944, when the escape corridor was closed for the last time, the Panzer regiment had no operational tanks.

Notes

1. The regiment's I.Abteilung was training in France but was far from combat ready and in fact would not join the division until 1945. In addition, the Panzerjäger battalion was still being formed and would take no part in the Normandy fighting.

Notes

1. Hauptsturmführer Karl Wilhelm Krause, the commander of the Flak-Zug, is credited with this innovation. The regiment's commander, Obersturmbannführer Max Wünsche, apparently used his influence with Hitler to have the project approved.

SS-Panzer-Regiment 12. Attached to 12.SS-Panzer-Division Hitlerjugend, this unit was almost continually engaged during the fighting in Normandy from the first day of the invasion.

Quite unofficially, and for the most part unreported, the regiment's II.Abteilung contained five companies in June 1944, numbered 5, 6, 7, 8 and 9. The exact distribution of the regiment's Pzkpfw IV tanks among the companies is a matter of some debate and is discussed in detail on page 63. In addition to the gun tanks the regiment had a number of field-modified anti-aircraft vehicles, possibly three, based on the chassis of a Pzkpfw IV and armed with four 2.0cm Flak 38 guns in place of the turret (1).

The division remained in the Caen area for much of the campaign, fighting against British and Canadian units and by the end of July operational tank strength had been reduced to twenty-two Panthers and thirty-nine Pzkpfw IV tanks with another twenty-three tanks of both type in short term repair. At that time 9.Kompanie was absorbed by 5.Kompanie and II.Abteilung reverted to the standard four-company organisation. In late August the division, with just ten operational tanks, was ordered to hold open the northern side of the gap Falaise and Argentan through which the remnants of 7.Armee were attempting to escape. On 8 September 1944 all surviving elements of the division, with the exception of a small Kampfgruppe formed around SS-Panzergrenadier-Regiment 26, were ordered back to Germany.

SS-Panzer-Abteilung 17. Attached to 17.SS-Panzergrenadier-Division Götz von Berlichingen the battalion had a full complement of forty-two Sturmgeschütz IV on hand in June 1944 and on 12 August received three Panzerbefehlswagen IV command tanks. The division took part in the fighting against the Americans around Carentan but was later surrounded at Coutances and only managed to break out after losing most of its equipment. The Pzkpfw IV command vehicles are only mentioned in the August report.

Below: Disabled near the village of Cheux, about 4 kilometres north-east of Villers-Bocage, in early July 1944 this Pzkpfw IV ausf H tank of 8.Kompanie, SS-Panzer-Regiment 12 was commanded by Untersturmführer Jeran. Note the unit insignia of 12.SS-Panzer-Division on the right side of the rear hull. This vehicle is also shown and discussed in the Camouflage & Markings section on page 24. At right: A Pzkpfw IV of 9.Kompanie. The unusual organisation of this regiment is examined on page 63.

1

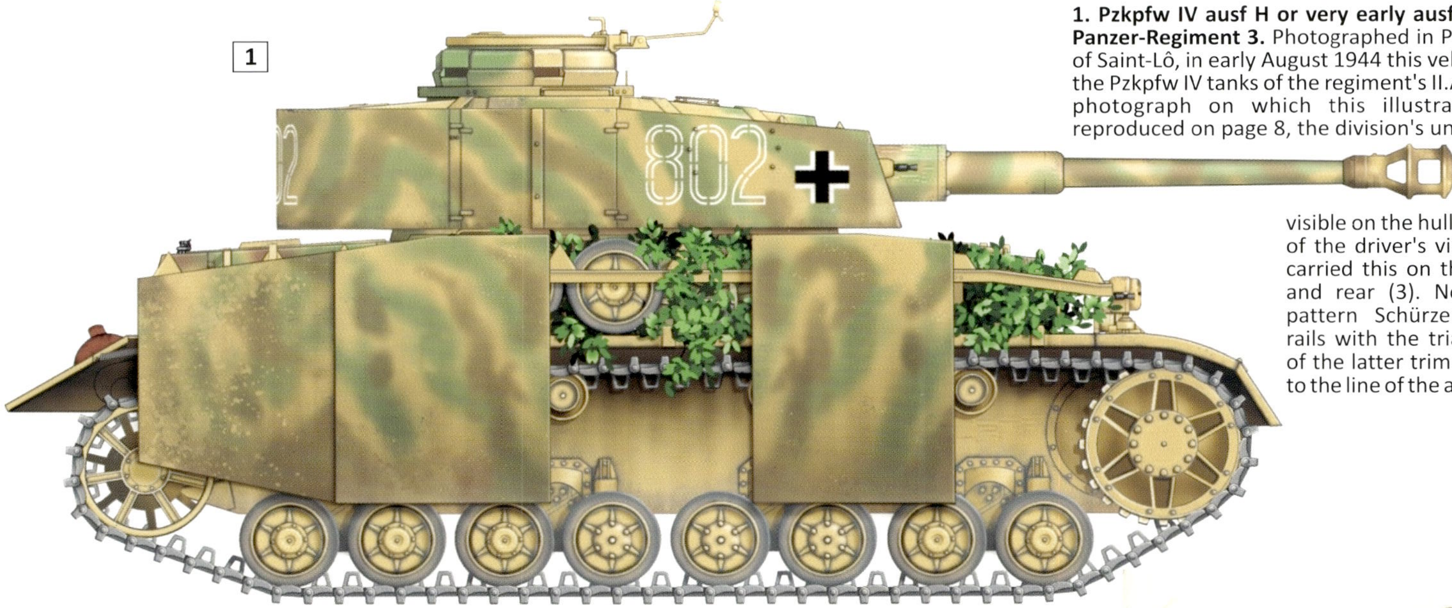

1. Pzkpfw IV ausf H or very early ausf J. 8.Kompanie, Panzer-Regiment 3. Photographed in Pont-Farcy, south of Saint-Lô, in early August 1944 this vehicle is typical of the Pzkpfw IV tanks of the regiment's II.Abteilung. In the photograph on which this illustration is based, reproduced on page 8, the division's unit insignia is just visible on the hull glacis to the left of the driver's visor. Most tank's carried this on the hull front (2) and rear (3). Note the second pattern Schürzen and support rails with the triangular hangers of the latter trimmed to conform to the line of the armour plates.

2

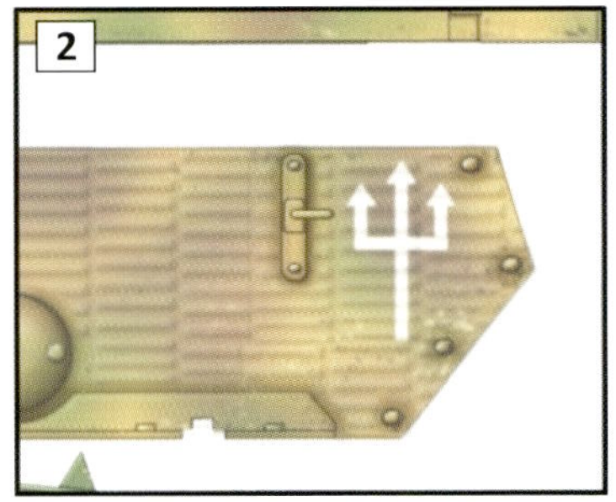

3

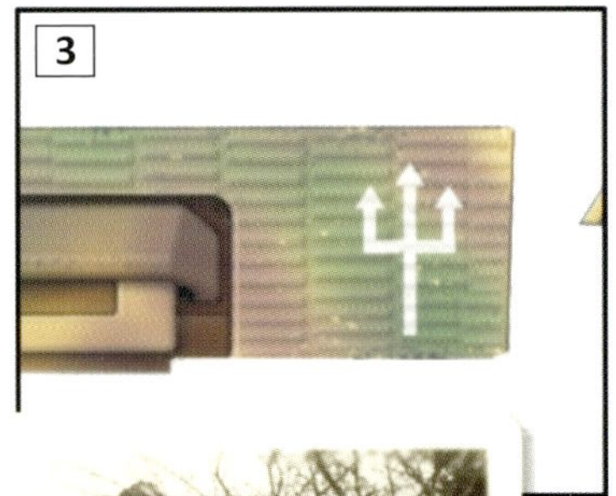

4

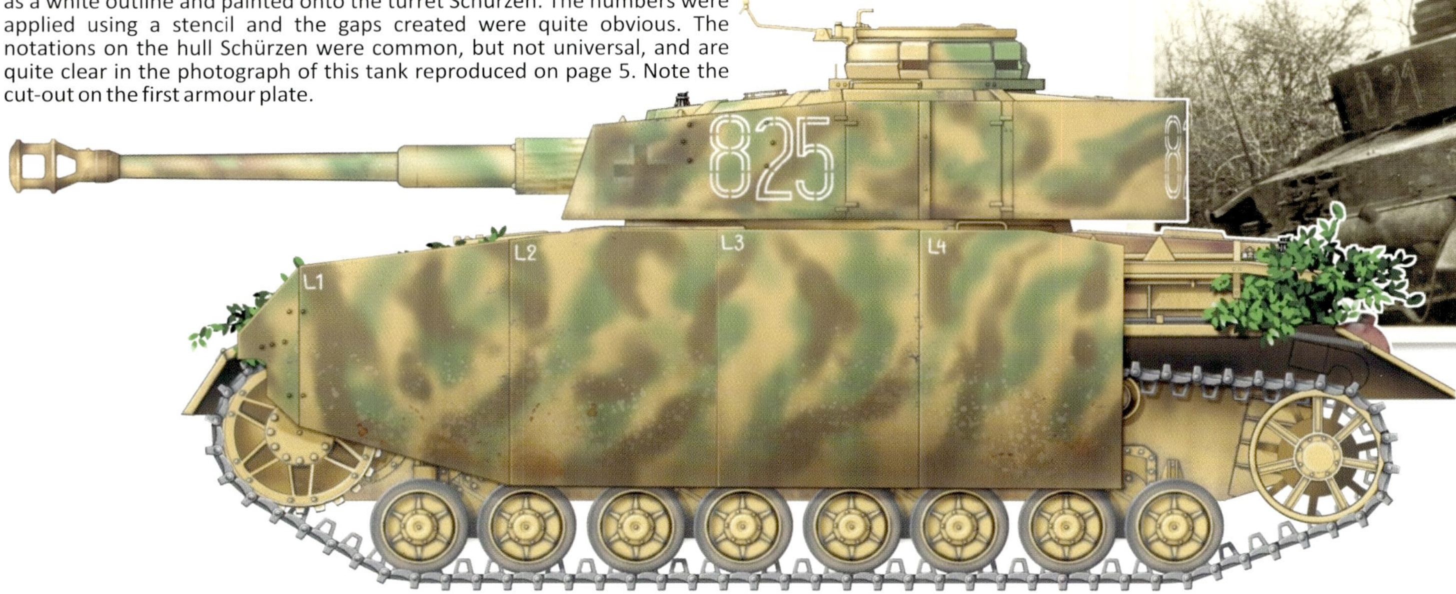

4. Pzkpfw IV ausf H. 8.Kompanie, Panzer-Regiment 3. All the tanks of the regiment's II.Abteilung were marked with large company numbers rendered as a white outline and painted onto the turret Schürzen. The numbers were applied using a stencil and the gaps created were quite obvious. The notations on the hull Schürzen were common, but not universal, and are quite clear in the photograph of this tank reproduced on page 5. Note the cut-out on the first armour plate.

Above: Pzkpfw IV ausf H Fgst. Nr. 86984 which is also shown in the photograph on page 1. Note the application of Zimmerit on the turret Schürzen, the company number repeated at the rear, the unit insignia of 2.Panzer-Division and the three spare track links indicative of Nibelungenwerk-built tanks.

1

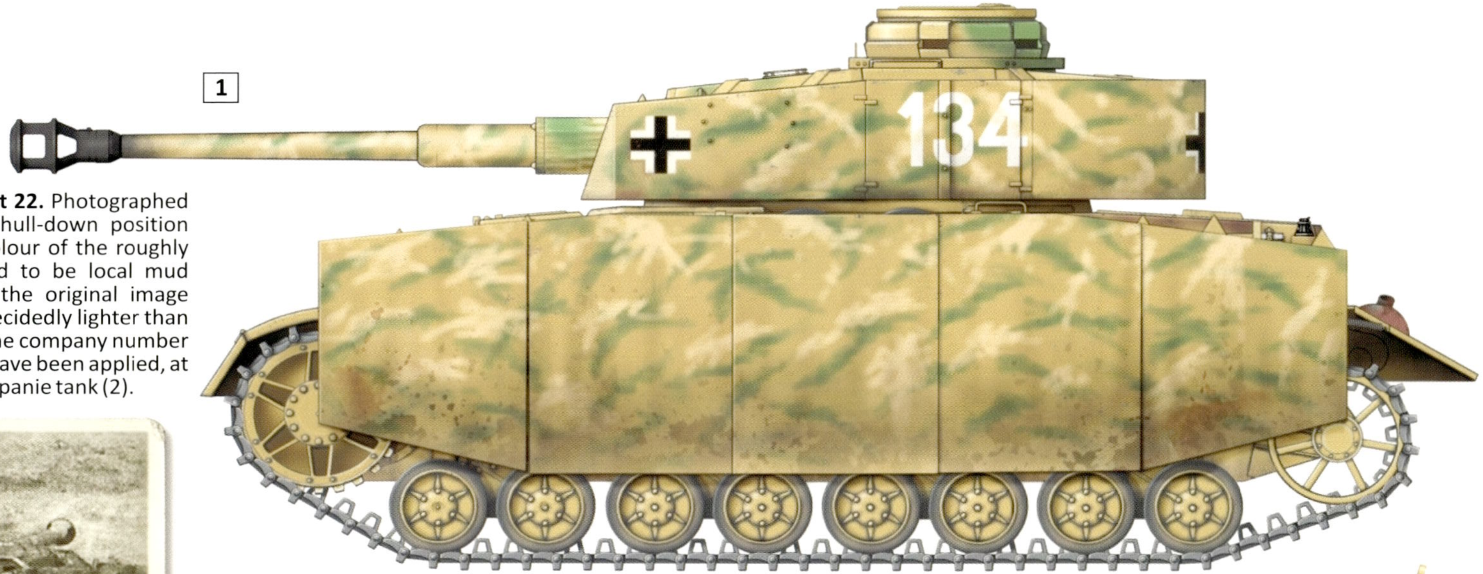

1. Pzkpfw IV ausf H. 1.Kompanie, Panzer-Regiment 22. Photographed on 13 July 1944, this tank was concealed in a hull-down position overlooking the Caen to Lébisey road. The pale colour of the roughly applied camouflage scheme is generally assumed to be local mud applied with a brush but close examination of the original image suggests that it may in fact be paint, although it is decidedly lighter than the Dunkelgelb base coat. The style used to depict the company number is typical of this regiment and the numbers seem to have been applied, at times, rather haphazardly as in the case of the 4.Kompanie tank (2).

2

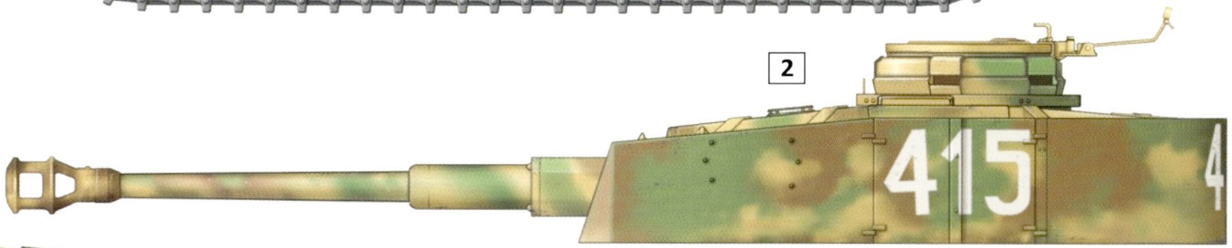

4

3

3. Pzkpfw IV ausf H. 8.Kompanie, Panzer-Regiment 22. Photographed at an Allied vehicle dump outside the village of Cambes-en-Plaine, north-east of Caen, after the Normandy battles this tank features the camouflage scheme of broad bands of Rotbraun and Olivgrün on the Dunkelgelb base seen on other later production Pzkpfw IV tanks of this regiment. (4). The company number is unfortunately obscured in our photograph and the rendering here is purely speculative, but entirely possible. The second pattern Schürzen rails have been reconstructed but parts are clearly visible.

1. Pzkpfw IV ausf H. 3.Kompanie, Panzerjäger-Lehr-Abteilung 130. Photographed at some time before May 1944 when four of the battalion's tanks were handed over to Panzer-Lehr-Regiment 130, this tank is fitted with the second pattern hull Schürzen and support rails that were introduced in September 1943. Note that the triangular hangers have not been trimmed as was common from October. The battalion's Pzkpfw IV tanks were all numbered as shown here and carried the tactical symbol denoting an anti-tank battalion with the company number and division's unit insignia on the front left fender (2).

3. Pzkpfw IV ausf H. 6.Kompanie, Panzer-Regiment 22. The four tanks handed over from Panzerjäger-Lehr-Abteilung 130 had their old numbers covered in a dark-coloured paint, all very roughly, and the new number added as shown here. This tank had previously been numbered 331. Note the first pattern hull Schürzen rails, side vision ports and bolted-on Zusatzpanzer armour visible in our photograph (4). Larger images of this tank are reproduced on pages 6 and 49.

1

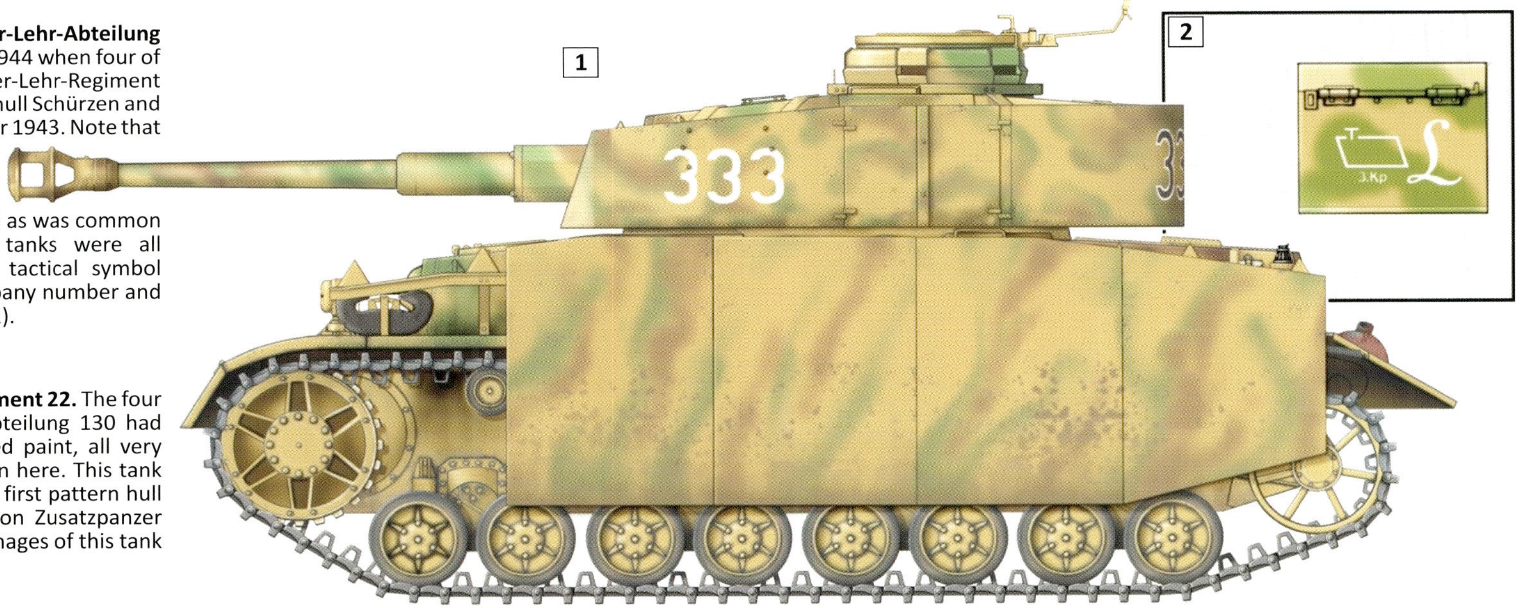

2

3

4

1. Pzkpfw IV ausf H. 5.Kompanie, Panzer-Lehr-Regiment 130. Photographed in the ruins of Saint-Gilles, west of Saint-Lô, following the Operation Cobra battles, this tank's company number has been applied directly over the loading label which was painted, in various styles, onto all German Army vehicles. A detailed view of the loading label is shown here (2) and indicates the vehicle type, weight and weight class, in this case 'S' for schwere ,or heavy.

1

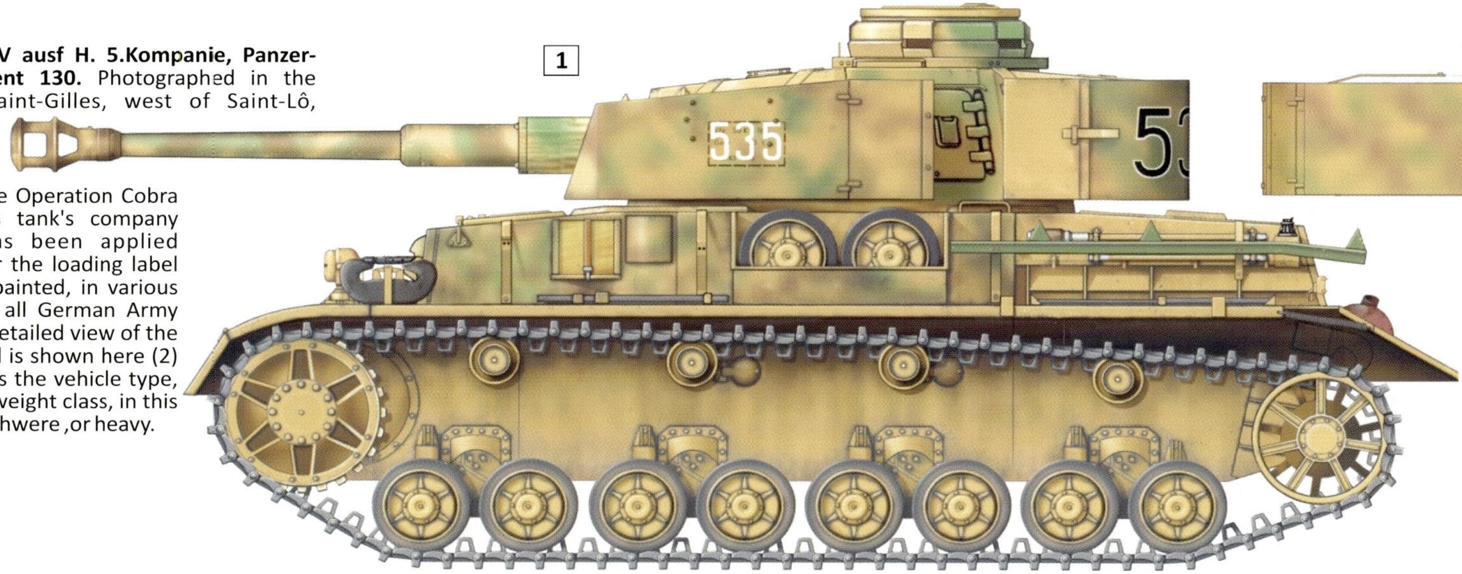

2

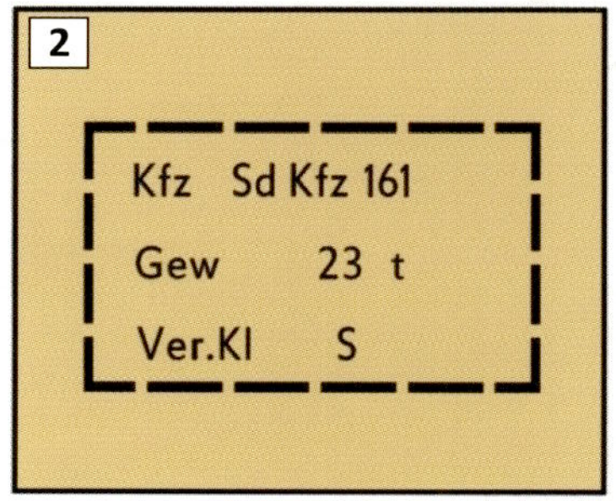

5

3. Pzkpfw IV ausf H. 6.Kompanie, Panzer-Lehr-Regiment 130. Knocked out in June 1944, during the fighting for Villers-Bocage, this tank was still in place when the Normandy battles ended. Note the second pattern hull Schürzen support rails, the forged hubcaps of the road wheels, cast rear idler and bracket on the rear hull side which held the long-handled spade facing forward, visible in our photograph (4). These features would suggest that this tank was assembled in late 1943. On 11 June 1944 the battalion commander, Prinz Wilhelm Schönburg-Waldenburg, was killed in action and from that day the tanks of II.Abteilung carried a rendition of his family crest (5) in the position shown here.

4

3

1. Pzkpfw IV ausf H. 7.Kompanie, Panzer-Lehr-Regiment 130. Although this tank was photographed (2) in Hungary just before the regiment left for the Western Front, it is safe to assume that its appearance would not have changed greatly when it arrived in Normandy. Note the very neatly applied turret number and tactical symbol identifying a Panzer battalion with the company number and division's unit insignia on the hull front (3).

1

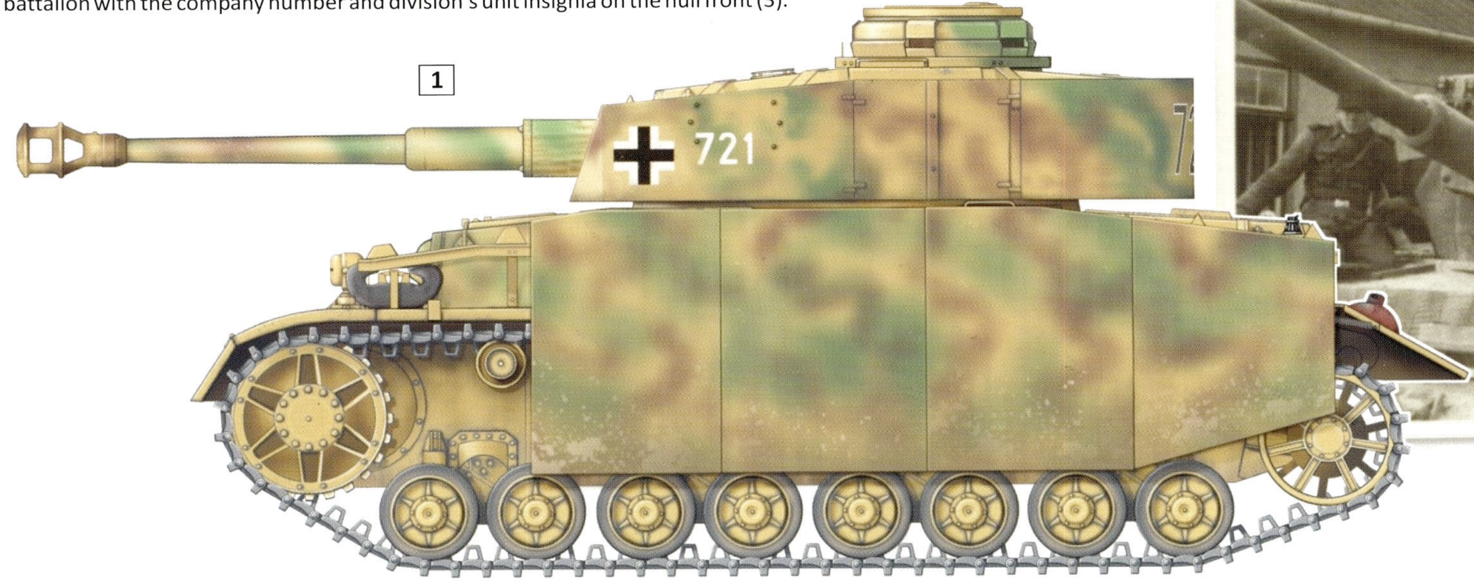

2

3

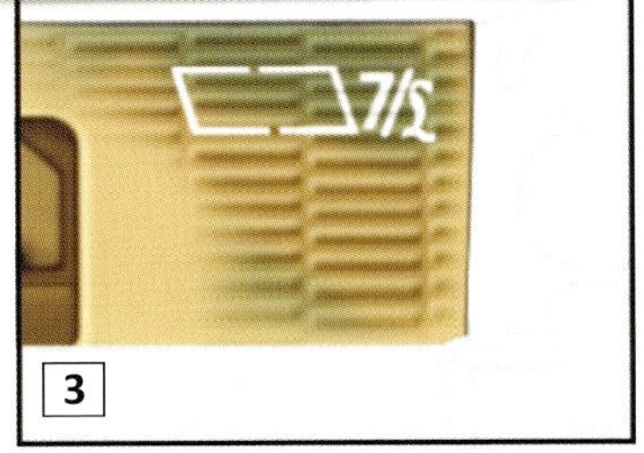

4

4. Pzkpfw IV ausf H. 8.Kompanie, Panzer-Lehr-Regiment 130. Photographed in Saint-Gilles, this vehicle was one of at least three Pzkpfw IV tanks, including number 535 shown on the previous page, that the regiment abandoned in the ruins of the village in late July 1944. A photograph of this tank is shown on page 12 and in the original image it is possible to see the turret traverse muffler which identifies the type. Note the roughly applied company number and Zimmerit on the hull side. The latter feature was not incorporated into production until early 1944, possibly in February. Note also that the supports for the hull Schürzen seem to have been completely removed although some of the brackets on the trackguard have been retained. The use of wire netting to hold foliage camouflage was common on all types of vehicles serving in western France at this time.

1

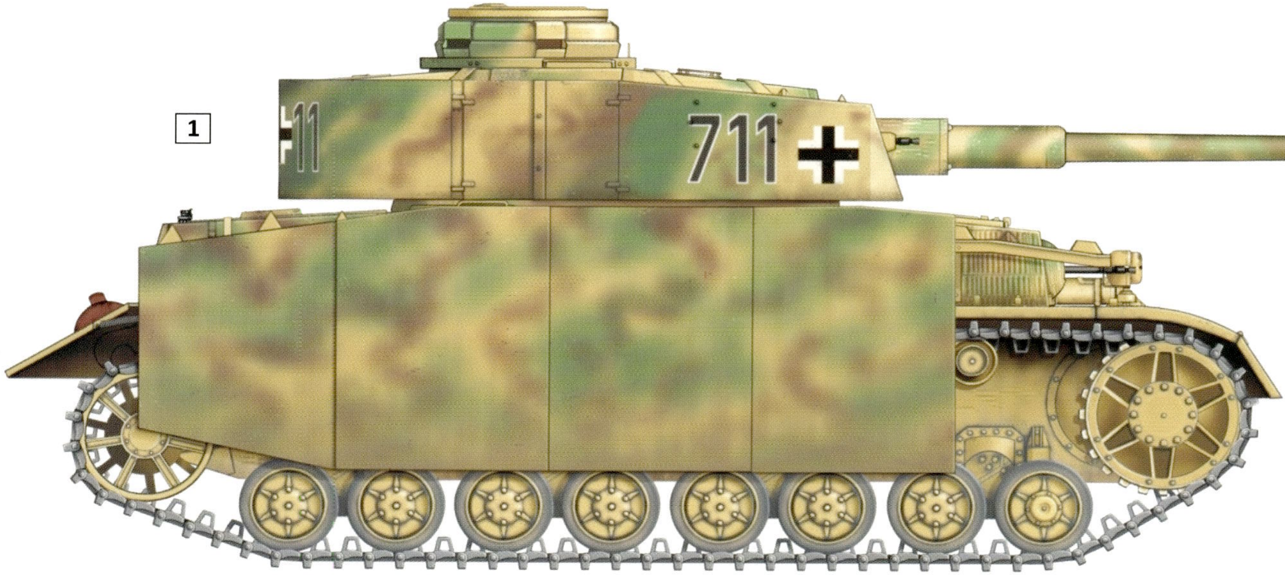

2

1. Pzkpfw IV ausf J. 7.Kompanie, SS-Panzer-Regiment 1. Commanded by Untersturmführer Gerhard Stiller, who led the company's first platoon throughout the Normandy battles, this tank survived the fighting in the Falaise Pocket only to be abandoned near Elbeuf-sur-Andelle, east of Rouen, due to a mechanical failure. The black company numbers, outlined in white, were common to most, but not all, the Pzkpfw IV tanks of this battalion as can be seen here (2). The photograph on which our rendering of tank 724 is based is somewhat unclear but it does seem that the number on the rear of the turret Schürzen is rendered as a white outline only. A photograph of Stiller's tank, before it was camouflaged and with what may be a red turret number, is shown on page 51.

4

3

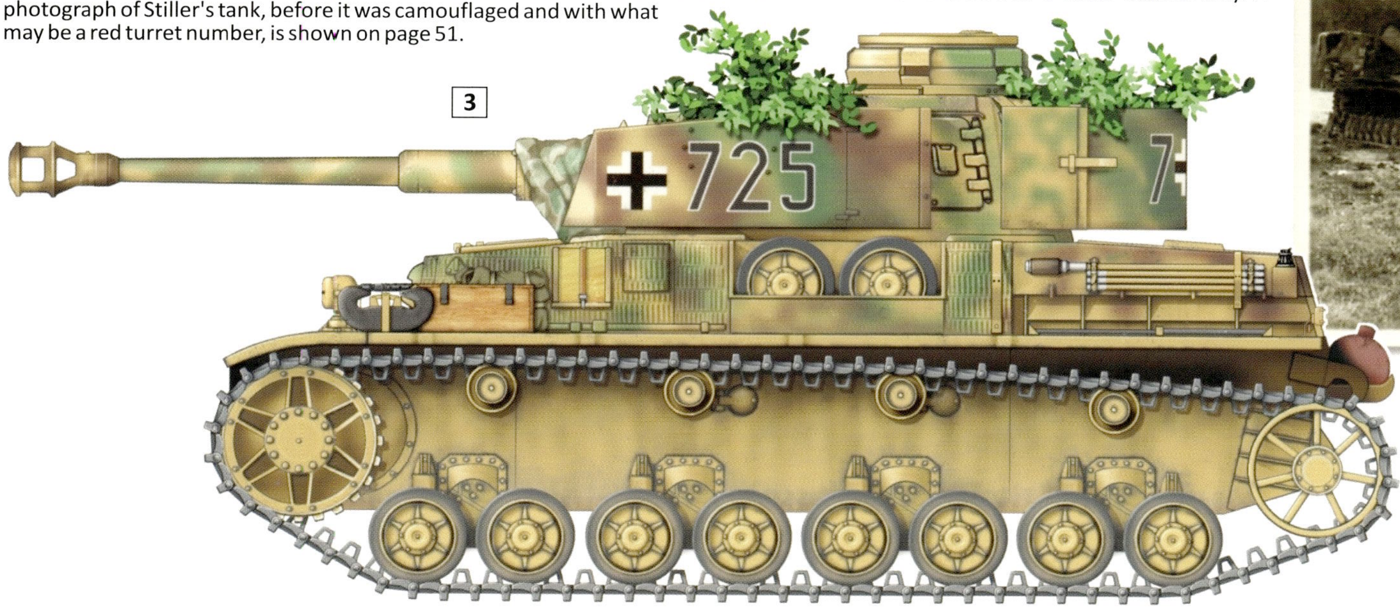

3. Pzkpfw IV ausf J. 7.Kompanie, SS-Panzer-Regiment 1. This tank, photographed while crossing one of the Seine River bridges, was one of the few survivors of the fighting in the Falaise Pocket. Photographs of the tanks of II.Abteilung taken in late July and early August 1944 show that the hull Schürzen were removed from many. Note the Zeltbahn tent quarter draped over the mantlet of the main gun. The method of displaying the company number on the rear of the turret Schürzen is shown in our photograph (4).

2

1

1. Pzkpfw IV ausf J. 6.Kompanie, SS-Panzer-Regiment 2. The use of a simple white outline to depict company numbers within the regiment's II.Abteilung seems to have been common, with a small number of exceptions, notably 7.Kompanie as mentioned below. The number was repeated on the rear of the turret Schürzen (2) as was the unit insignia of 2.SS-Panzer-Division (3) and (4). A photograph of this 6.Kompanie tank is reproduced at the top of page 51.

3

4

5. Pzkpfw IV ausf H. 7.Kompanie, SS-Panzer-Regiment 2. Although the numbering system used by this regiment is a matter of some debate it seems fairly certain that the use of the number 9 denotes a battalion or company headquarters vehicle. This tank was probably commanded by Obersturmbannführer Karl Kloskowski, who led 7.Kompanie during the Normandy battles, and contrary to what has been written elsewhere it was not a purpose-built Befehlspanzerwagen command tank but a field modification. Many of the company's tanks had rather hastily applied turret numbers, as can be seen from our photograph (6), and some of these may have been the tanks handed over by Panzer-Regiment 15 in early August 1944.

5

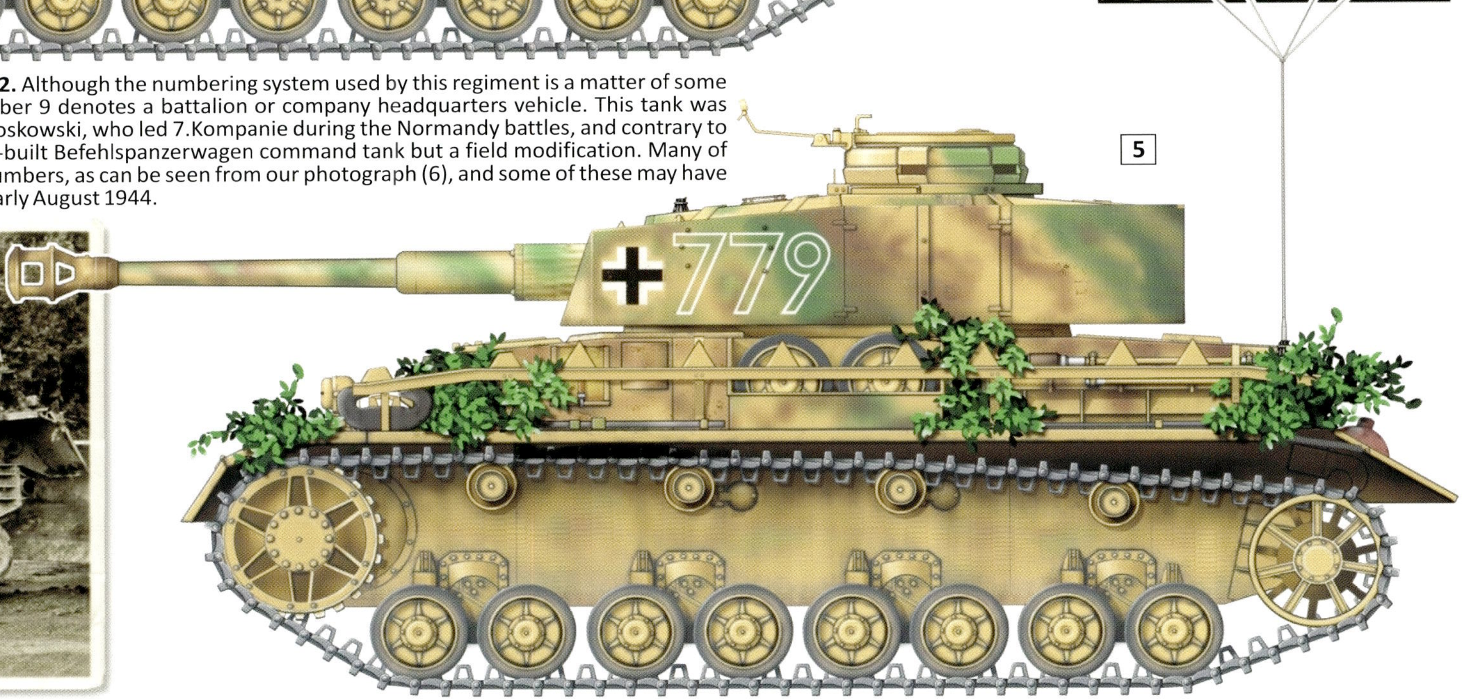

6

1. Pzkpfw IV ausf H. 5.Kompanie, SS-Panzer-Regiment 12. Commanded by Unterscharführer Willy Kretschmar, the hull Schürzen plates and support rails of this tank were badly damaged after the fighting for Hill 112 near Caen, as can be seen from our photograph (2). The hull Schürzen are of the second pattern and some of the plates, but not all, were coated with Zimmerit anti-magnetic mine paste although it is notably absent from the turret armour. As with most tanks of this battalion the turret number has been rather roughly repainted by hand, probably after the reorganisation that is examined in detail on page 63. Note the kill rings on the barrel which are noticeably absent in earlier images. A photograph of this tank is shown on page 4.

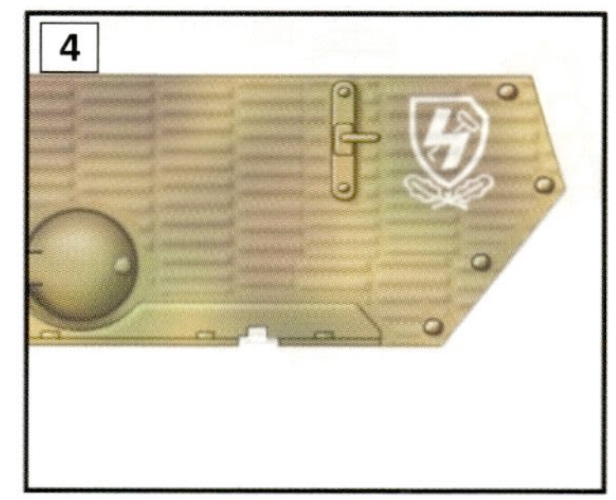

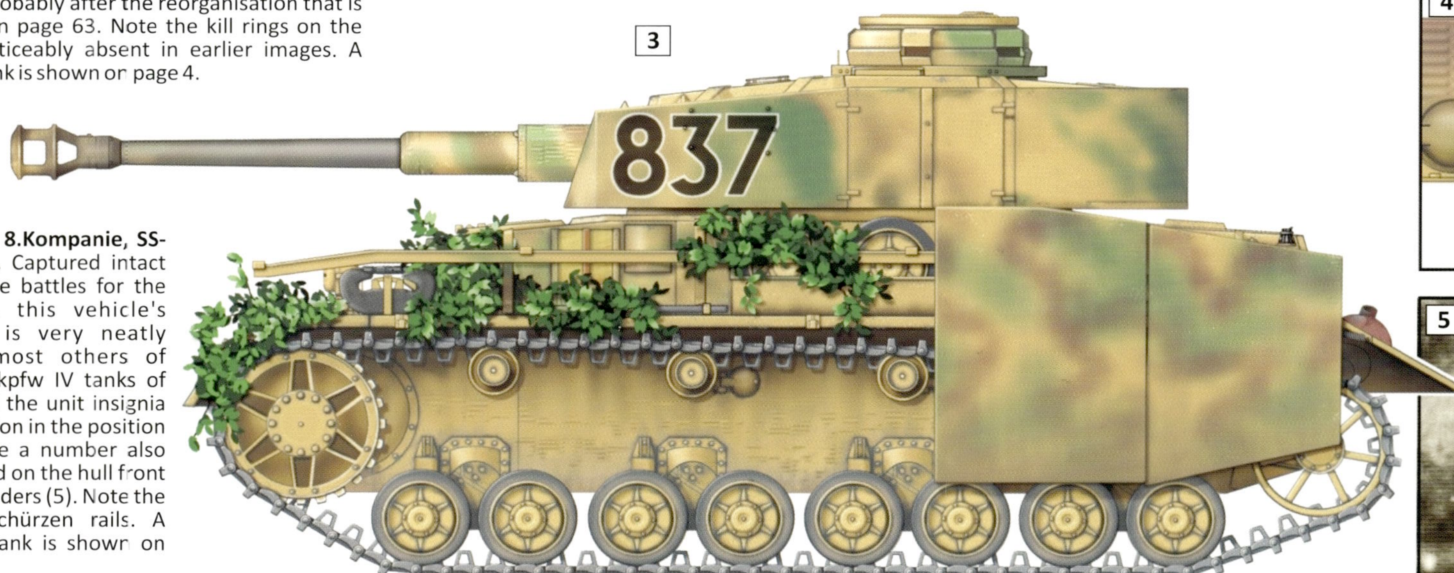

3. Pzkpfw IV ausf H. 8.Kompanie, SS-Panzer-Regiment 12. Captured intact near Cheux, after the battles for the Carpiquet airfield, this vehicle's company number is very neatly executed, unlike most others of I.Abteilung. Most Pzkpfw IV tanks of this regiment carried the unit insignia of 12.SS-Panzer-Division in the position shown here (4) while a number also had the badge painted on the hull front or one of the front fenders (5). Note the first pattern hull Schürzen rails. A photograph of this tank is shown on page 16

1. Pzkpfw IV ausf H. 6.Kompanie, SS-Panzer-Regiment 12. Photographed while the regiment was training in Belgium during the spring of 1944, this vehicle is fitted with the second pattern hull Schürzen introduced in September 1943 with the distinctive cut-out in the first plate that was a feature of Nibelungenwerk-built tanks.

1

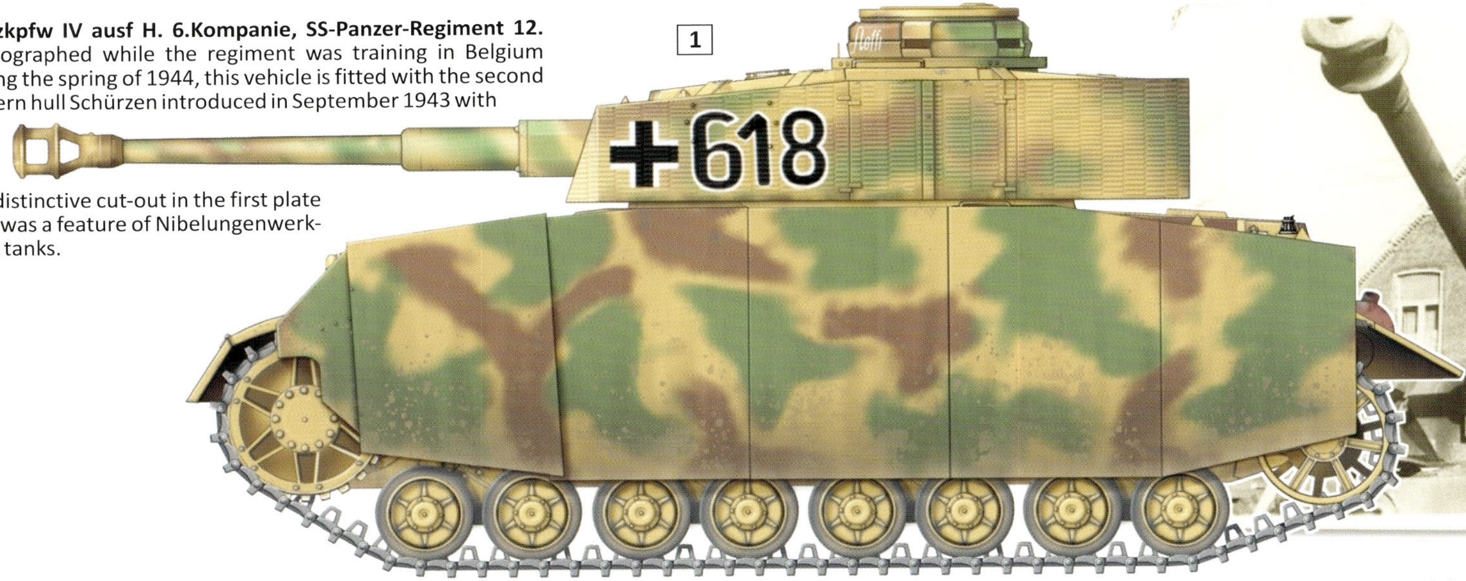

2

3

Above: Note the name 'Steffi' painted onto the commander's cupola which can be clearly seen in our photograph (2) and is shown in an enlarged view here (3). The use of female names was common in this battalion with some tanks having more than one, as shown here (4).

4

5. Pzkpfw IV ausf H. 6.Kompanie, SS-Panzer-Regiment 12. This tank, Fgst Nr.86934, was assembled by Nibelungenwerk in November 1943 and later captured in running order and evaluated at the British School of Tank Technology. It has the later cast return roller and rear drive sprocket and the hull was built without the side visors. Note the field-modified bullet splash guard mounted on the turret roof (6). The notation 'braun ark' on the barrel of the 7.5cm gun refers to the oil used in the recoil mechanism. In moderate temperatures Bremsflussingkeit Braun was used and for intense cold Bremsflussingkeit Arktik was employed. When both were mixed in equal parts it was referred to as Braun Ark.

5

6

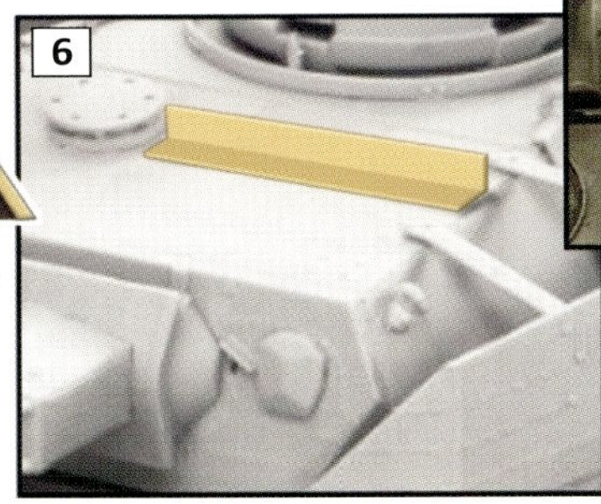

1. Pzkpfw IV ausf J. 4.Kompanie, Panzer-Regiment 33. This tank took part in the fighting around the small town of Marolles-les-Braults, south-east of Alençon, in early August 1944. The company numbers were applied throughout the battalion in the colour and style shown here, the battalion staff rendered as I01, I02 etc. The camouflage scheme made up of broad bands of colour was common in the regiment and is also seen on a number of

1

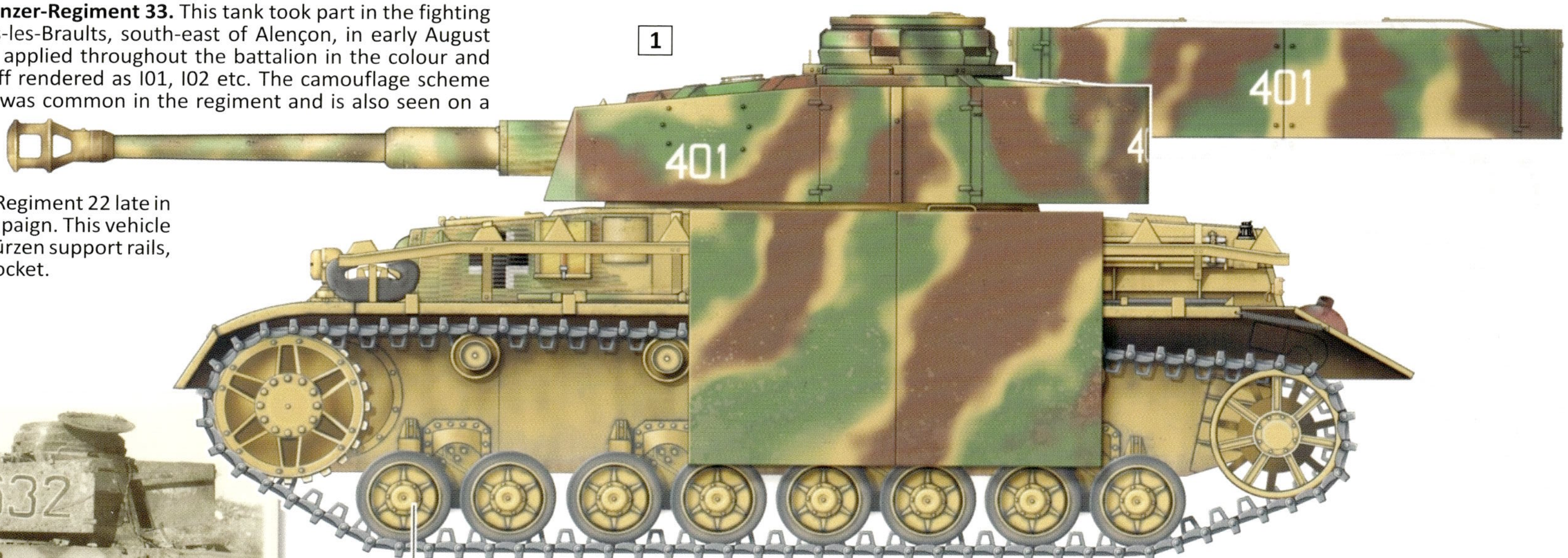

Pzkpfw IV tanks allocated to Panzer-Regiment 22 late in the last weeks of the Normandy campaign. This vehicle is fitted with the second pattern Schürzen support rails, later cast return rollers and drive sprocket.

3

2. Pzkpfw IV ausf G. 6.Kompanie, Panzer-Regiment 15. Disabled and abandoned near Lyon in August 1944, this early model tank has the folding antenna on the hull side, early drive sprocket and rear idler, bolted-on Zusatzpanzer armour on the hull glacis and Nebelwurfgerät smoke candle dischargers on the turret. The latter were dropped from production in June 1943 and were supposed to have been removed by units in the field shortly afterwards. Company numbers were displayed on the turret Schürzen but the style varied greatly, even within the same company, and a variation is shown in our photographs (3) and (4).

2

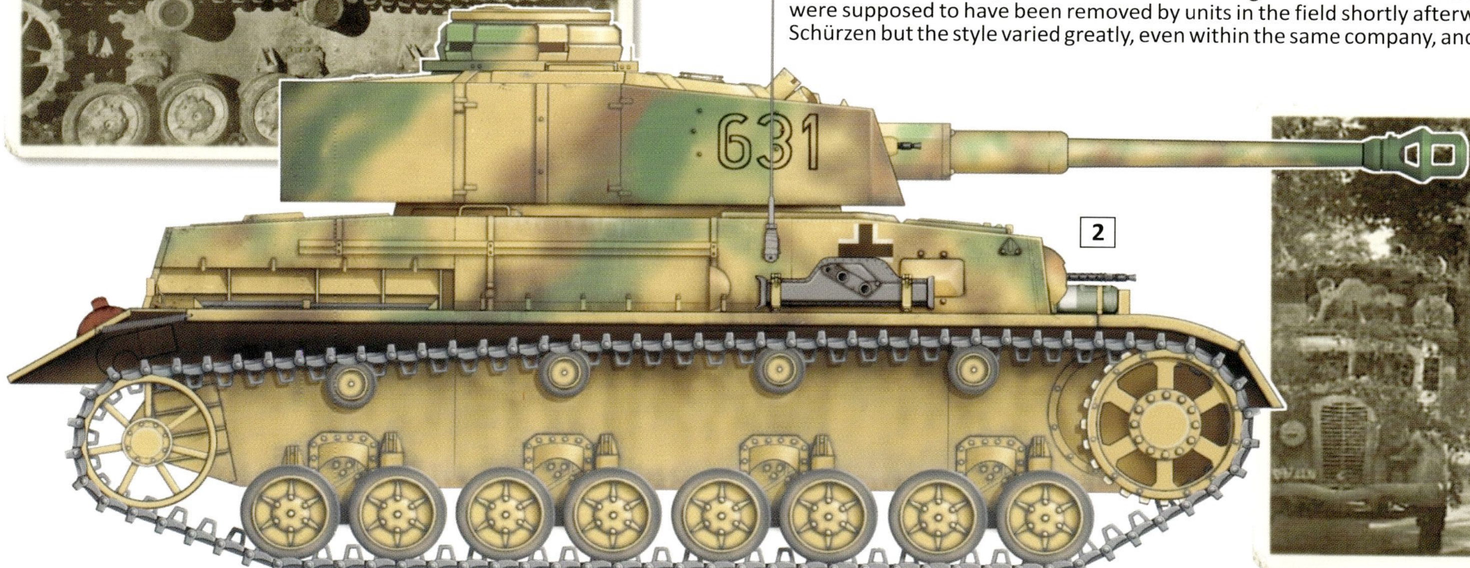

4

PANZER IV

SS-PANZER-REGIMENT 12

JAMES NIGEL PANGANIBAN

1/35 SCALE

Based on the 1/35 scale Pzkpfw IV ausf H kit from Academy Plastic Model, which is marketed as a mid-production model, James' replica depicts a tank of 6.Kompanie, SS-Panzer-Regiment 12. This vehicle is also shown and discussed on page 25 of the Camouflage & Markings section of this book. The Zimmerit texture was supplied with the kit as an extra and is produced by DEF.Model, a Korean company which offers a comprehensive range of accessories and figures.

Another view of James' completed model. Note the mix of early and later track types, a common feature, and the cut-out in the first Schürzen plate. The latter was an identifying feature of Nibelungenwerk-built tanks.

Below: Details of James' Pzkpfw IV ausf H model including (1) the exhaust, power traverse muffler and spare tracks. The rear of the turret Schürzen with the company number and the engine deck (2). The Bosch headlight, fire extinguisher and C-shaped towing hooks (3). Note the texture of the trackguard. (4). The commander's cupola with circular splash guard and anti-aircraft machine gun bracket.

Although the original kit was first released in 1984, the model has been upgraded with new parts several times, the last as recently as 2020, and James found the detail and ease of assembly to be generally good.

James added many small details including the weld beads around the turret. The camouflage scheme, typical of summer 1944, was created using Tamiya Dark Yellow, Dark Green and Red Brown, all from the acrylic range. Chips and scratches in the finished paintwork were added by hand and artist's oils were used to replicate streaking, dust and dirt. Weathering pigments and tufts of grass were added to the tracks.

PANZER IV

SS-PANZER-REGIMENT 12

THEODOROS KALAMATAS

1/72 SCALE

Theodoros' replica is based on the Zvezda 1/72 scale kit which he found to be well detailed and accurate. The Zimmerit texture is a photo-etched product from Part, a Polish company which specialises in armour model accessories but as it is specifically designed for the Revell kit a certain amount of cutting and pasting was necessary to achieve the correct effect.

Once built, the model was painted with acrylic paints from Tamiya and Mr Hobby and finished with a variety of the weathering products offered by AMMO by Mig Jimenez and AK Interactive. The foliage is produced by Bear's Scale Modelling (BSM), a relatively new company from Poland run by modeller Miroslav Serba. Theodoros' model can be purchased and he is available for commission work. His contact details are given on *page 64.*

PANZER IV

PANZER-REGIMENT 22

YVAN GARNIER

1/35 SCALE

Yvan built his early model vehicle based on Dragon Models' 1/35 scale Pzkpfw IV ausf C kit. Some additional small features, such as handles and latches, were created from plastic sheet and metal rod.

Below: The interior of the turret showing the extensive detailing which can still be glimpsed through the side access hatch

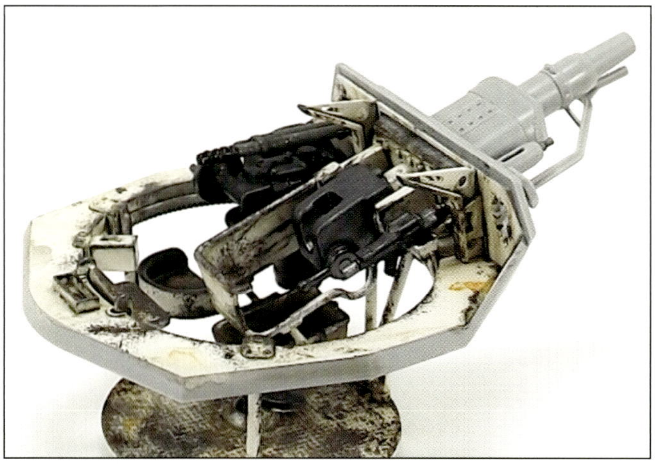

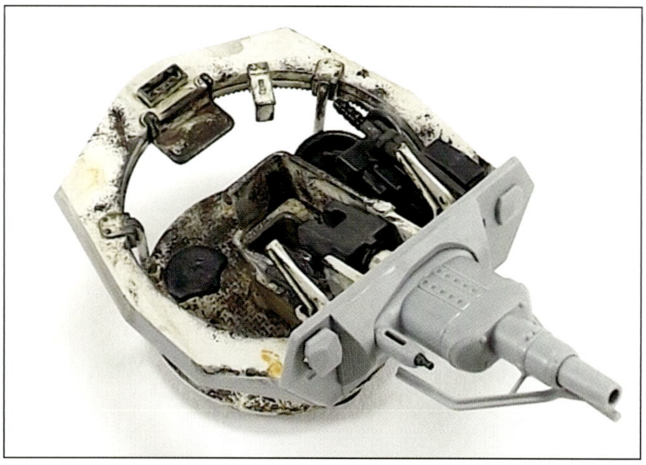

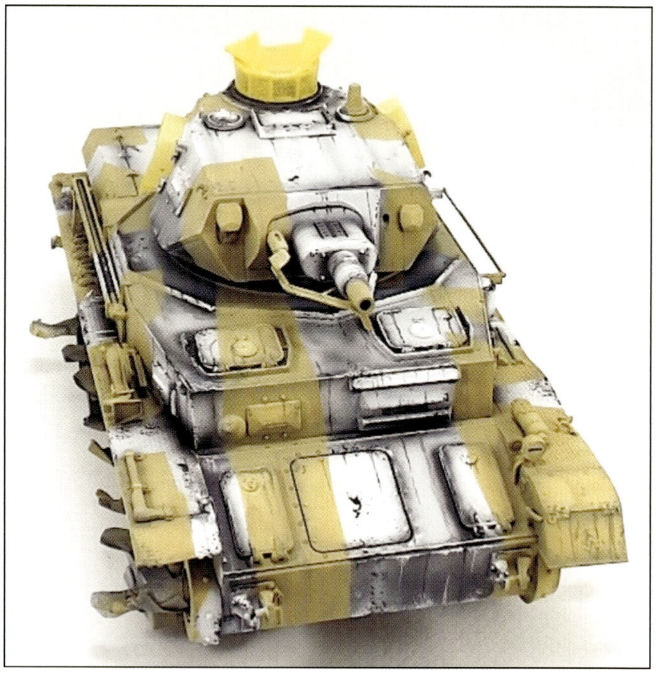

The partly-built model was first painted with Tamiya acrylic colours and the final camouflage scheme was achieved with a mix of Model Colour paints from AV Vallejo. The unusual pattern is based on an actual vehicle of 8.Kompanie, Panzer-Regiment 22 photographed on 25 August 1944.

The completed model, ready to be fixed to its base. After painting the camouflage scheme Yvan weathered the finished replica with filters and washes from AK Interactive and AMMO by Mig Jimenez. Note the damage to the left side trackguard and the mix of first-pattern and cast road wheel hubcaps, both of which are present on the full size version abandoned by its crew after the Normandy battles.

Another view of Yvan's completed model showing the early commander's cupola, side-mounted antenna and its protective wooden trough and the engine deck.

The completed model on its scenic base. The crew figures are from MK35, a French company which offers an extensive range of accessories, and Yvan built the base up with plaster filler. It was then painted in earth tones and subjected to the pigments that he had used to weather the tank. The hedgerow in the background is a mix of small trees from Model Scene and the foliage is made up of various products from Plus Model and MiniNatur, a German firm that specialises in model railway scenic accessories.

PANZER IV

SS-PANZER-REGIMENT 12

KONSTANTINOS KALOGEROPOULOS

1/48 SCALE

Konstantinos chose Tamiya's 1/48 scale Pzkpfw IV ausf H kit as the basis of his model. The Zimmerit texture is also from Tamiya and was reworked with a Dremel motor tool to reproduce the effects of battle damage. Brackets were added to the hull Schürzen which were made from sheet styrene using the kit offerings as a template. The Schürzen rails were also modified to represent the second pattern versions. A number of small details were added including the wire of the Bosch headlight and the handles and latches of the engine deck, which is quite plain on the original kit. The tool clamps are from the photo-etched brass detail set produced by Aber. The completed model was then primed with Gunze 1500 Mr Finisher surfacer gray. The images at the bottom of the page show Konstantinos' model after it was sprayed with MRP paints to replicate the factory primer red colour and the Zimmerit, which is a rather bland beige shade. The camouflage scheme was applied using Tamiya's new lacquer Dark Yellow, Dark Green and Rotbraun mixed with NATO Brown.

The kit was then treated with thin layers of Scratch Effects from AMMO by Mig Jimenez and, once dry, the camouflage scheme was painted using Tamiya acrylics. Scratches and chips in the paintwork were achieved with the help of some worn brushes and an old airbrush needle.

A fine sanding stick also helped remove parts of the colour from the Zimmerit surfaces in random spots. The markings are from Star Decals, which this author helped to design, and Microset and Microsol solutions from Microscale helped them to conform them to the Zimmerit texture. Once the transfers were set the model was weathered with Flory Models dark dirt wash and dedicated streaking oil paints from AMMO.

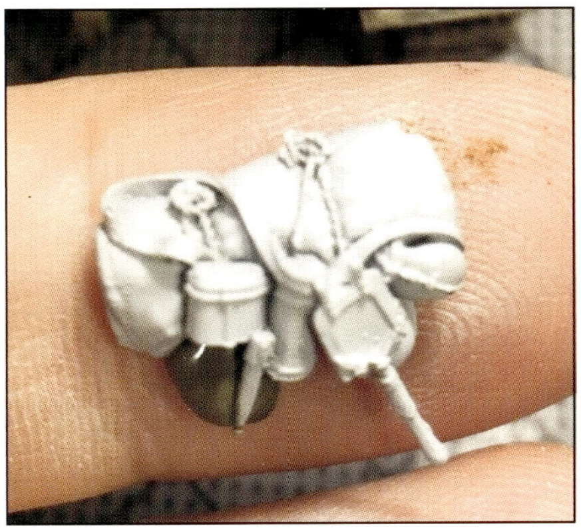

A view of the finished model showing the extremely realistic foliage camouflage made by Dioramapresepe, an Italian firm which offers an extremely wide range of scale greenery and terrain effects. A number of accessories, including the two jerry cans, bucket and army field gear are from Tamiya and Black Dog, a Polish company which has been featured several times in this series.

Below: Another view of Konstantinos' completed model showing the turret roof detail as well as the extensive reworking of the hull Schürzen rails. Above: The markings from Star Decals on the turret Schürzen and commander's cupola show how well they conform to surface texture. Johan Lexell, who runs the company, is sensitive to the needs of modellers of German armour and has worked for years to perfect this aspect of his transfers and also their accuracy.

The Panzerkampfwagen IV was numerically the most important tank in service with the German army during the Second World War and it should come as no surprise that scale replicas of the various marques, from the Pzkpfw IV ausf A of 1937 to the final version of the Pzkpfw IV ausf J, are available to the modeller. These vary in size from 1/144 scale miniatures to 1/16 scale motorised models and include highly detailed and accurate replicas as well as basic, snap-together kits aimed at the wargames market. To list all these would not be possible in the space available and I have limited this section to the major manufacturers of 1/35 scale and 1/48 scale kits and to the versions of the tank which were most commonly encountered during the time period covered by this book, that is to say the Pzkpfw IV ausf H and the early examples of the Pzkpfw IV ausf J. Similarly, I have only made brief mention of the smaller, specialised companies which produce additional items in resin, photo-etched metal and other materials that allow a level of super-detailing which is almost limitless. Hopefully this volume will be the first in a series of books covering the complex and fascinating story of the Panzerkampfwagen IV and these aspects will be examined in greater detail in future titles. I should mention that I have limited any comment or critique of these kits or products to those of which I have first-hand knowledge. An index of manufacturers can be found on page 64.

TAMIYA INC

This Japanese company has been manufacturing plastic kits since 1959, releasing their first armour model two years later. By the early 1970s Tamiya was producing a large range of armoured vehicles and accessories and the company was almost solely responsible for the rise in popularity of 1/35 scale. Tamiya's early kits were, for their day, highly detailed, relatively easy to assemble and, most importantly, affordable. The company's 1/35 scale Pzkpfw IV ausf H was first released in 1975 but has been upgraded since then and the Pzkpfw IV ausf J is a completely new kit. From 2003 Tamiya began releasing a series of armour models in 1/48 scale and the current catalogue includes a late production Pzkpfw IV ausf H and an early production Pzkpfw IV ausf J. The Trumpeter 1/16 scale Pzkpfw IV is a re-boxing of the motorised Tamiya kit that was originally released in 2008.

Above, left: Tamiya's 1/48 scale Pzkpfw IV ausf H built with the photo-etched brass Zimmerit set from Eduard Model Accessories. Above, right: A detail of the early production Pzkpfw IV ausf J in 1/48 scale. At right: The early production Pzkpfw IV ausf H model with Zusatzpanzer bolted-on hull armour.

DRAGON MODELS LTD

At the time of writing this Hong Kong-based company's catalogue contained over thirty models of the Pzkpfw IV in 1/35 and 1/72 scale including vehicles which were in service during the summer of 1944. It should be noted that a number of these are repackaged kits, some containing additional parts or markings, and a number are pre-painted 1/72 scale display models. Nevertheless, this is an impressive total and the Dragon kits of the Pzkpfw IV ausf H and early model Pzkpfw IV ausf J are highly detailed replicas containing optional parts, photo-etched elements and a choice of markings. To complement the vehicles there is a large range of highly detailed, plastic multi-pose figures.

Above, and at right: Dragon Models' Pzkpfw IV ausf H later production versions with Zimmerit in both 1/72 and 1/35 scale. Numerous marking options are contained in these kits including those for SS-Panzer-Regiment 12 shown here.

Below: The 1/35 scale Pzkpfw IV ausf H model completed with the photo-etched brass detail set from E. T. Model faeturing the second pattern hull Schürzen rails. Details of the same kit, built straight from the box, are shown at the bottom of the page.

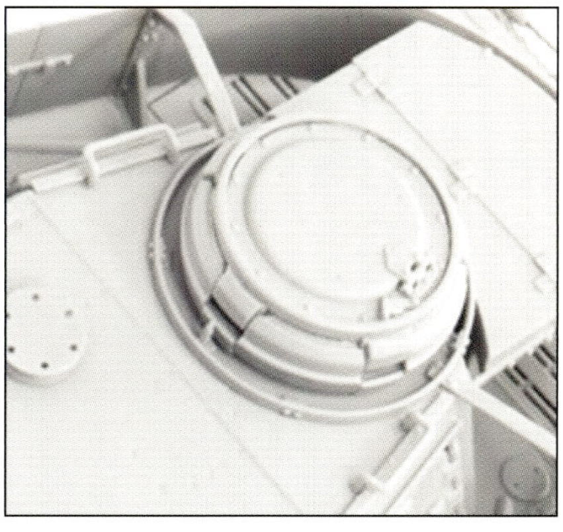

TRUMPETER PLASTIC MODEL KITS/HOBBY BOSS

Often thought to be separate entities, these Chinese companies are in fact owned by the same corporation, are located at the same address and are essentially the same firm although different models are marketed separately under the two labels. In 2015 Trumpeter released a 1/16 scale Pzkpfw IV ausf H and an early and later version of the Pzkpfw IV ausf J, both of which are shown below. Hobby Boss produces a series of early Pzkpfw IV tanks armed with the short barrel 7.5cm KwK L/24 gun, all in 1/35 scale, which are re-boxed versions of the 2005 kits from Tristar.

MINIART CREATIVE MODELS

Based in Ukraine, this company has been producing 1/35 vehicles, figures and aircraft models since 2001. The current catalogue contains a number of models based on the Pzkpfw IV ausf G, Pzkpfw ausf H and Pzkpfw ausf J and all include photo-etched brass parts and extensive marking options. The Krupp-Grusonwerk kit shown below contains photo-etched parts, individual track links, with a jig to help with assembly, and extensive interior detailing.

Above: Two of the 1/35 scale Pzkpfw IV kits from Miniart that are relevant to the summer 1944 period.

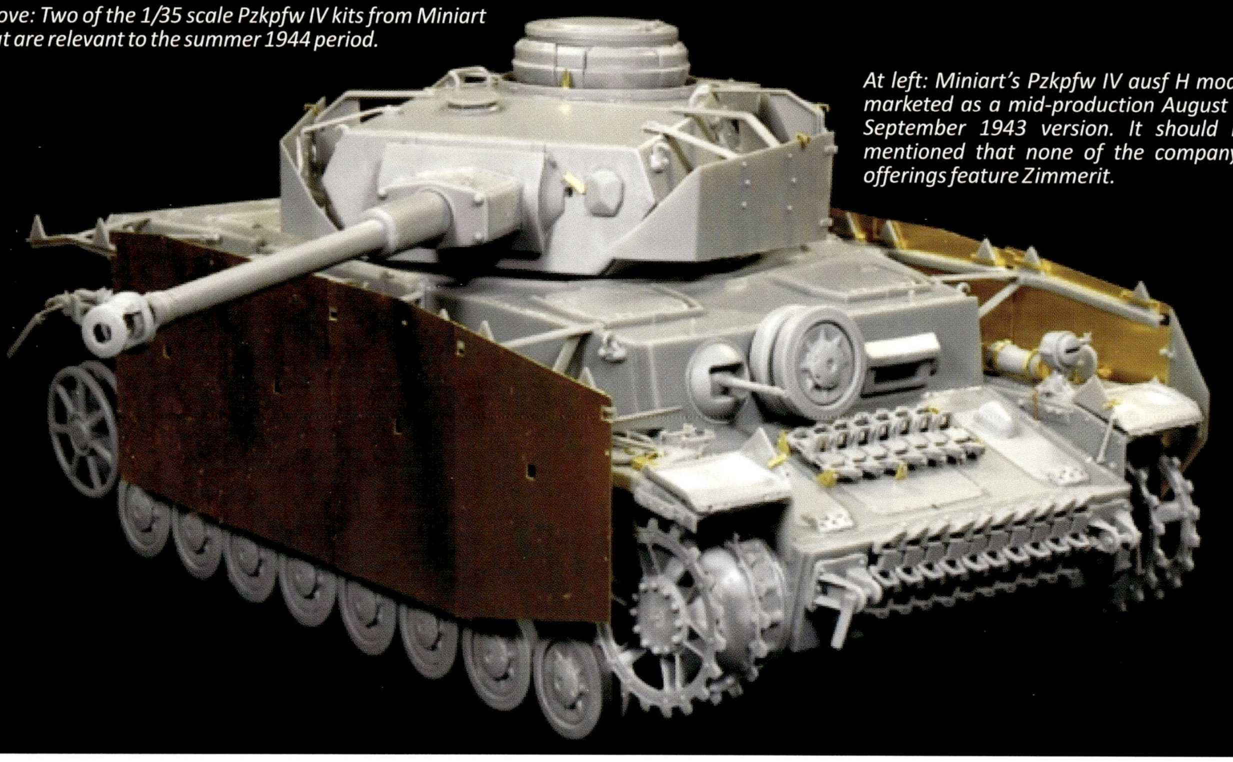

At left: Miniart's Pzkpfw IV ausf H model marketed as a mid-production August to September 1943 version. It should be mentioned that none of the company's offerings feature Zimmerit.

ACADEMY PLASTIC MODEL

The company's Pzkpfw IV ausf H was originally released in 1994 and over the next twenty years was re-boxed and re-released several times, also under the Minicraft label. The latest version is a completely new kit and includes optional Zimmerit texture and although the tracks are the single-length rubber variety they are reasonably well detailed. The three marking options all represent tanks of SS-Panzer-Regiment 12 as they appeared in the summer of 1944 during the battles in Normandy. The Pzkpfw IV ausf H kit released by Airfix in 2018 and the 2019 offering from Revell are the same model with different markings. An early production Pzkpfw IV ausf J was to be released in 2020 but is not included in the company's updated catalogue.

At left: Details of Academy's Pzkpfw IV ausf H kit including the optional Zimmerit texture produced by DEF Models.

The Academy Pzkpfw IV ausf H model built and painted following the kit instructions.

RYE FIELD MODEL

A relatively new model company based in Hong Kong, Rye Field Model's current catalogue includes an early production Pzkpfw IV ausf H and a Pzkpfw ausf J which could be built as a vehicle which participated in the Normandy campaign. The latter includes extensive interior detailing but neither contain the option of Zimmerit texture which would be appropriate for a mid-1944 tank. A Pzkpfw IV ausf G and a Pzkpfw IV ausf G/H were to be released in 2021 but do not appear to be available at the time of writing. The company also produces sets of workable tracks.

At left and below: Rye Field Model's 1/35 scale Pzkpfw IV ausf H. The hull Schürzen and their rails are the first series pattern and with the addition of Zimmerit, and the omission of the field-modified metal plate on the turret roof, this tank would be typical of the vehicles that fought in Normandy.

ITALERI

This Italian company has been producing plastic models since its foundation in 1962 and the company's 1/35 scale Pzkpfw IV ausf H kit is an old model that has been re-boxed under various labels including the now-defunct Testors and Tomy, as well as Zvezda and Revell. The latest incarnations were re-released in 2012 and 2012 with new parts including photo-etched details and a metal gun barrel. The 1/72 scale offering is the ESCI model first released in 1974 and then again in 1988 with individual track links. Although old, this is a reasonably good kit and it should be said that ESCI models were somewhat groundbreaking for their day. A number of Italeri's armour models have also been released under the World of Tanks label and also as part of the Girls und Panzer franchise.

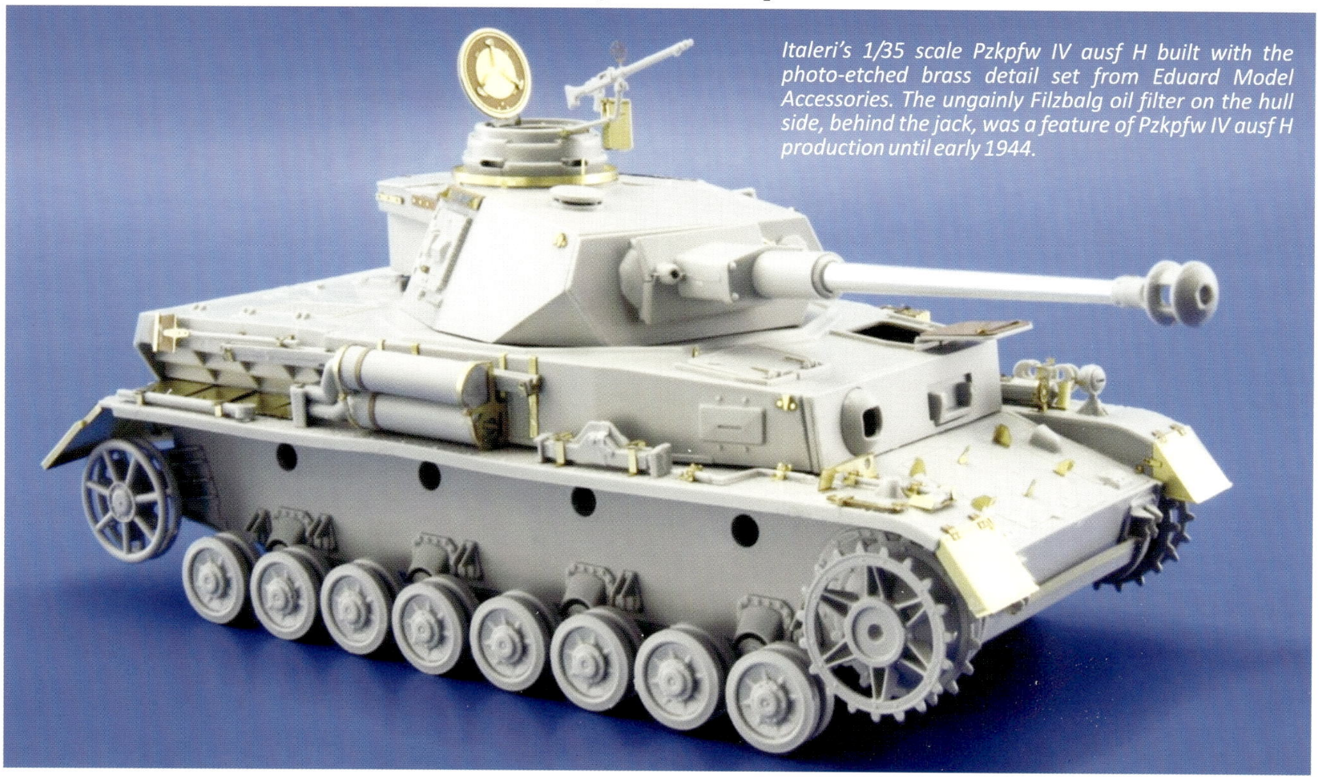

Italeri's 1/35 scale Pzkpfw IV ausf H built with the photo-etched brass detail set from Eduard Model Accessories. The ungainly Filzbalg oil filter on the hull side, behind the jack, was a feature of Pzkpfw IV ausf H production until early 1944.

BORDER MODEL

Situated just outside Shanghai, this company began releasing a kits based on the Pzkpfw IV chassis and hull in 2019, the latest offering being a reworked 1/35 scale Möbelwagen 3.7cm Flakpanzer. Border Model also manufactures workable tracks and photo-etched brass detail sets for their replicas. The Pzkpfw IV ausf H model, the most relevant for our period, contains crew figures, photo-etched parts, a metal gun barrel, a tool to inscribe the Zimmerit texture and fourteen marking options, six of which depict tanks which served in France in 1944.

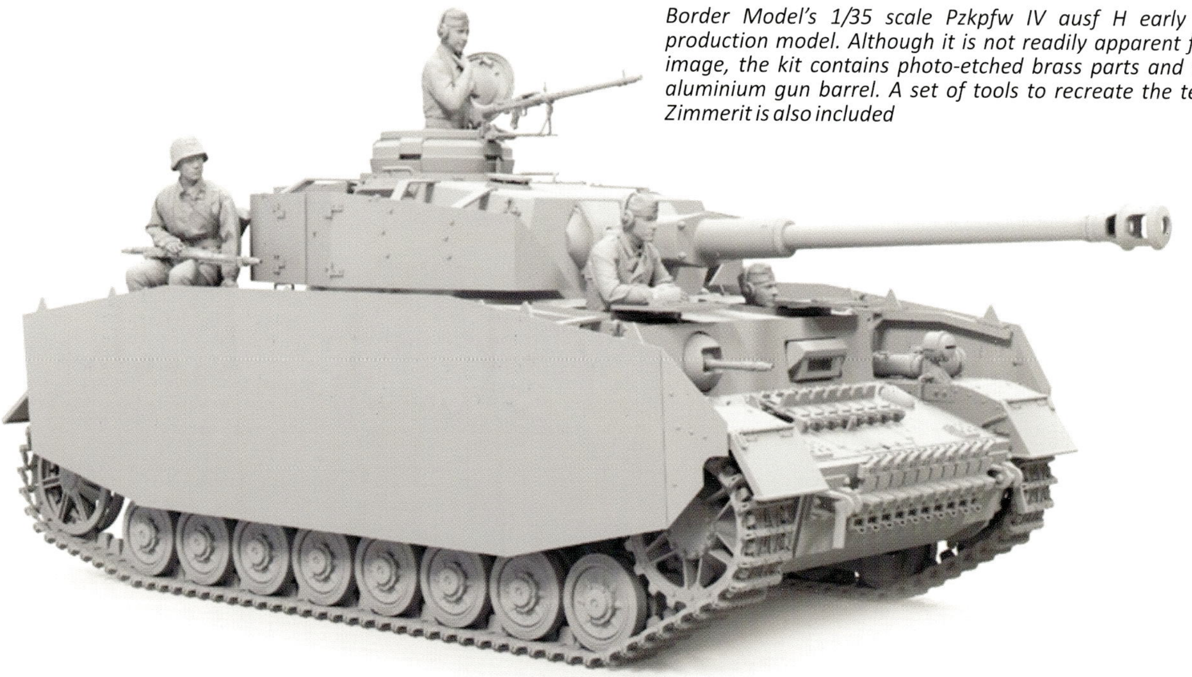

Border Model's 1/35 scale Pzkpfw IV ausf H early or mid-production model. Although it is not readily apparent from this image, the kit contains photo-etched brass parts and a turned aluminium gun barrel. A set of tools to recreate the texture of Zimmerit is also included

The development of the Panzerkampfwagen (Pzkpfw) IV tank, from its introduction as the Begleitwagen in 1937 to the last production vehicles which left the assembly lines in April 1945, is far too complex a subject to cover in a single volume. In this section I have concentrated on the Pzkpfw IV ausführung (ausf) H, which was the most common type encountered on the Western Front during the Normandy battles, and the early examples of the Pzkpfw IV ausf J, which was entering service in mid-1944. This is not to say that earlier models were not on hand with front-line units at that time and some of these are shown in the photographs reproduced in this book and also the Camouflage & Markings section. The technical details of those versions will be explored in future volumes. Production and allocation figures are shown on pages 57 and 58 and it should be mentioned that these two categories are not intended to correspond, but merely show the activity that took place in a given month. The shipment of tanks to operational units was the responsibility of the Heereszeugamt and could take place many weeks after the assembled vehicles were accepted by the Heereswaffenamt.

The Pzkpfw IV ausf H was assembled by the firms of Friedrich Krupp AG, Vogtländische Maschinenfabrik (Vomag) and at the massive Nibelungenwerk (NiWerk) at Sankt Valentin in Austria which produced more tanks than both its competitors combined. The first examples left the assembly lines in May 1943 and the early production models of the new tank were externally similar to its predecessor, the Pzkpfw IV ausf G. The distinctive welded or bolted-on armour, which had been gradually introduced between May 1942 and January 1943, continued to be fitted to the front of the hull and fighting compartment. The main armament was the 7.5cm Kwk 40 L/48 gun, which had also been incorporated into the production of the Pzkpfw IV ausf G from April 1943, replacing the slightly shorter 7.5cm L/43 version. A redesigned muzzle brake was introduced at the same time. Protective hull and turret Schürzen, or skirts, were standard. The brackets used by Krupp and Vomag to hold the turret Schürzen in place were slightly different to those employed by NiWerk. The visors on the hull sides were dropped from production and the turret roof was strengthened with the armour plate of the forward section increased from 10mm to 16mm thickness while the rear section, around the commander's cupola, was increased to 25mm. The radio antenna was moved from its folding position on the hull right side to the left rear side where the insulator was held in a metal bracket. The wooden trough which protected the

Photographed in July 1944 this Pzkpfw IV ausf H of 6.Kompanie, Panzer-Regiment 22 was built by Vomag between June and August 1943, before the introduction of Zimmerit. Points to note are: A. 30mm bolted-on armour on the hull front. B. One-piece 80mm glacis. C. Rubber-tyred return rollers. D. Welded drive sprocket. E. Early cast hubcaps. F. First pattern Schürzen rail, badly damaged. G. Bosch headlight. H. Hull side vision ports. I. The late pattern double-circular baffle muzzle brake which is most often seen on Pzkpfw IV ausf J models. Another photograph of this tank is reproduced on page 6.

This Pzkpfw IV ausf H was assembled by Nibelungenwerk in August or September 1943 and lacks a coat of Zimmerit anti-magnetic mine paste. Note the support brackets (A) for the hull Schürzen rail, which are of the second type. The vehicle in the background has the support rail for the first pattern hull Schurzen (B). Although difficult to see here, the second bracket of the turret Schürzen (C) is the type used only by Nibelungenwerk. Other points to note are: D. The rubber-tyred return rollers. E. Cast hub caps. F. C-shaped towing hooks. Note also the round hole access point (G) on the track guard brace, to the left of the Kugelblende machine-gun mount, which was another identifying feature of this manufacturer. The armour plate on the turret roof is a field modification.

Notes

1. The Panther I is more commonly referred to today as the Panther ausf D and the rather complicated history of the naming of these tanks is explained in detail in *TankCraft 34: Panther Medium Tank, German Army and Waffen-SS, Eastern Front Summer 1943.*

folded antenna was removed from the hull side and its place taken by a Filzbalg oil-filter and for some time a wooden strip was fitted to the lower edge of the rear portion of the turret Schürzen, although it was now superfluous. Hulls which had been built to accommodate the folding antenna featured a metal plug in its place but these are difficult to see in contemporary photographs. This modification was also carried out on the last of the Pzkpfw IV ausf G models.

The Pzkpfw IV ausf H was fitted with a reinforced final drive and this necessitated a redesigned, welded drive sprocket which was fitted to all vehicles except the first thirty tanks assembled by Vomag which had the earlier version. One distinguishing feature of NiWerk-produced tanks was the three spare track links held on the hull rear plate on the left side.

All vehicles were allocated a Fahrgestelle, or chassis, number abbreviated here as Fgst. Nr. The major modifications incorporated into the production of the Pzkpfw IV ausf H are listed below.

April 1943. Before assembly of the Pzkpfw IV ausf H could begin the Krupp plant was ordered to abandon its manufacture and to prepare for production of the Panther I or Panther II (1) which were then in the development stage. In Pzkpfw IV ausf G production, which continued into May 1943, the longer 7.5cm L/48 gun replaced the L/43 version and a redesigned multi-baffle muzzle brake was introduced. As a defence against Soviet anti-tank rifles, armoured skirts, referred to as Schürzen, were fitted to the turret and hull. The turret Schürzen of NiWerk-assembled tanks were set 50mm further forward.

May 1943. In the final month of Pzkpfw IV ausf G production a Filzbalg oil filter was mounted on the right side track guard replacing the flexible antenna which was moved to the left side of the rear hull. The antenna could still be folded downwards on these tanks to lie across the engine deck.

The Filzbalg filters were standard fittings on the Pzkpfw IV ausf H but the antenna and its insulator were held in a metal bracket in a static position on the hull left side at the rear, above the air-intake. The wooden trough on the hull side which had protected the folding antenna of the Pzkpfw IV ausf G was left in place and held a spare aerial.

The first Pzkpfw IV ausf H left the assembly plants in this month. For the entire project Fgst. Nr. were allocated in blocks to the three major manufacturers in the range 84401 to 90600. During May the managers at Krupp were notified that the Panther II project would be terminated and that some Panther I should be completed as Bergepanther recovery tanks.

.........continued on page 52

This Pzkpfw IV ausf J of 6.Kompanie, SS-Panzer-Regiment 2 was abandoned north of Saint-Lô on 9 July 1944. Clearly visible here are: A. The second pattern Schürzen rails with their triangular hangers. B. The Schürzen rail support brackets. C. The track guard brackets which held the Schürzen plates at an inclined angle. D. Convoy light. E. Nibelungenwerk trailer towing bracket. F. Cast steel return rollers. G. Later pattern rear idler. H. FuG 5 radio antenna and insulator. The unit insignia of 2.SS-Panzer-Division can just be seen on the rear hull (I).

An early production Pzkpfw IV ausf J of 7.Kompanie, SS-Panzer-Regiment 1 photographed in May 1944 while the regiment was refitting in Belgium. This tank is fitted with a full set of turret and hull Schürzen and this image shows how the triangular hangers, here trimmed, were fitted into the U-shaped brackets welded to the inside face of the armour plates.

..........continued from page 50

June 1943. Nebelwurfgerät, smoke candle dischargers, were no longer fitted to the turret sides (1) as it was believed that small arms fire could cause them to prematurely ignite. A bracket for an anti-aircraft MG 34 machine gun was fitted to the commander's cupola.

Beginning in June the hull side vision ports were to be partially dropped from production and completely phased out by September. Presumably the hulls already built with the necessary apertures were to be fitted with vision ports.

Zusatzpanzer, or additional armour, in the shape of 30mm thick plates, had been welded or bolted to the hull glacis and superstructure front in the Pzkpfw IV ausf G production run from May 1942, gradually becoming standard in February 1943. The welded plates were dropped from the manufacture of the Pzkpfw IV ausf H at the end of May in favour of the bolted-on type and both were superseded by a homogenous 80mm-thick armour plate in June 1943. This is, however, rather simplistic and vehicles could be seen with a mix of all three.

August 1943. The side visors and rear pistol ports on the turret were dropped from production. Krupp was ordered to halt all Panther I assembly and begin production of the Pzkpfw IV, with a monthly target set at 150. Later the same month the target was dropped to 100.

September 1943. As an economy measure, cast bump stops replaced the welded brackets of the road wheel arms and forged hubcaps were introduced in place of the cast versions. Late in the month the factories began applying Zimmerit anti-magnetic mine paste to the vertical armour surfaces. For a short period Zimmerit was also applied to the hull and turret Schürzen and the horizontal surfaces around the driver's and radio operator's hatches. In the same month rectangular splash guards were introduced for those hatches.

October 1943. The rubber-tyred return rollers were replaced by cast steel versions. A cast metal rear idler replaced the welded, tubular version that had been standard since 1941 with the manufacture of the Pzkpfw IV ausf F. Both

Notes

1. Photographic evidence shows that these were indeed included in early Pzkpfw IV ausf H production.

Below: Production features of the Pzkpfw IV ausf G and Pzkpfw IV ausf H. A. The 30mm Zusatzpanzer additional armour fitted to the glacis. Note the bolts and the way the plate is cut to accommodate the Kugelblende machine-gun mount. B. The Filzbalg oil filter. C. Nebelwurfgerät, or smoke candle dischargers. A small number of Pzkpfw IV ausf H tanks were fitted with these before they were dropped from production in June 1943. D. The pivoting radio antenna and insulator of the Pzkpfw IV ausf G. The antenna folded down and towards the rear and was held in the wooden trough visible here. E. The muffler for the internal auxiliary power traverse engine. F. The position of the antenna and its insulator on the Pzkpfw IV ausf H. Note the three spare track links, a feature of Nibelungenwerk-built tanks.

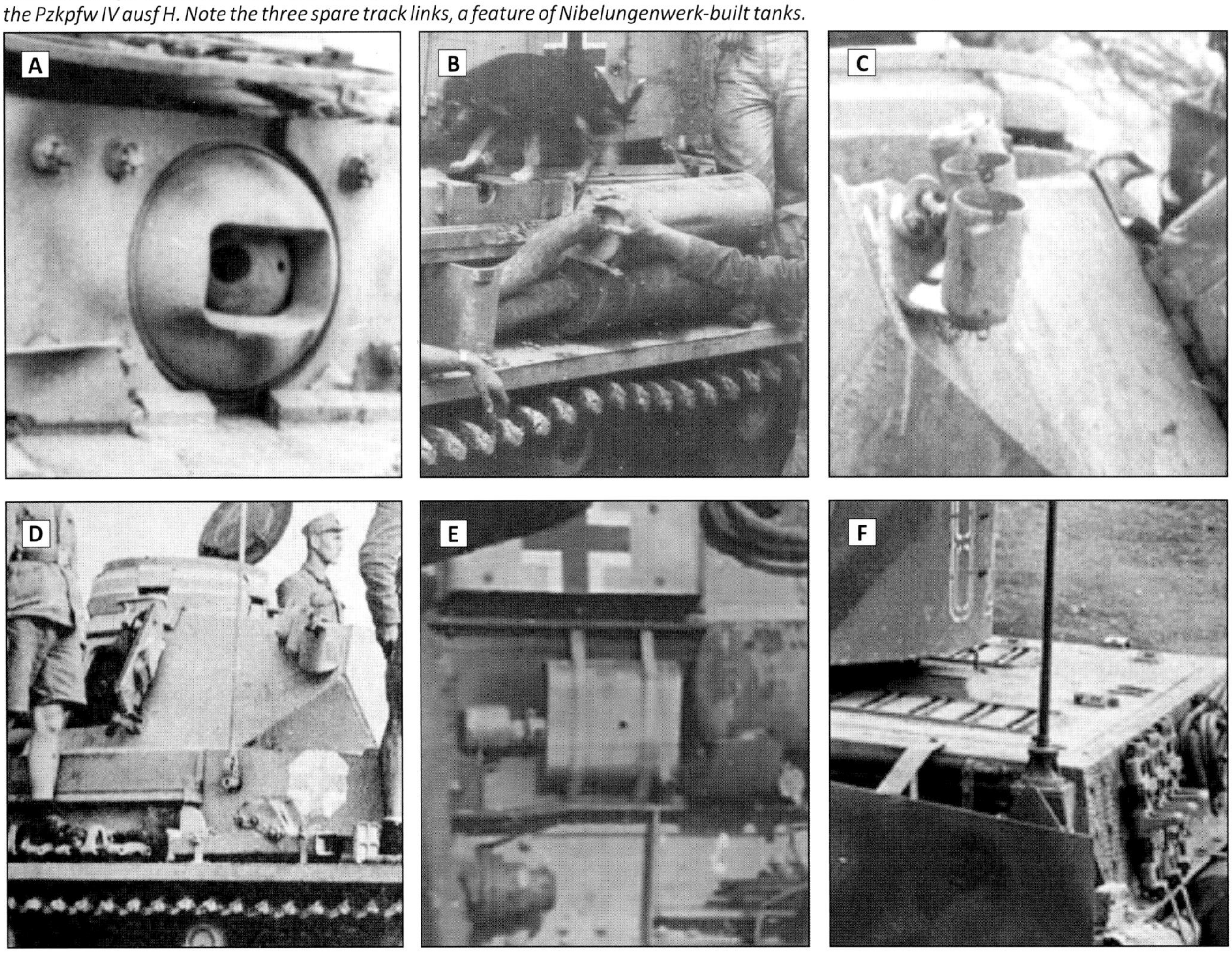

Although this Pzkpfw IV ausf H was photographed on the Eastern Front I have included the image here as it clearly shows the turret Schürzen support brackets used by Vomag and Krupp (A). Note also the bolted-on Zusatzpanzer armour of the glacis (B) and hull front (C). The hull still has the side vision ports (D) and the early rear idler (E).

these changes were introduced gradually and Krupp in particular was slow to fit the new return rollers and rear idler. A redesigned rail to hold the hull Schürzen was introduced. The rails were now fitted with triangular holders or hangers and these were slotted into brackets welded to the inside face of the Schürzen. It is common to see the earlier type of hull Schürzen plates re-worked to fit the new rails.

November 1943. Wooden planks were bolted between the track guards and hull Schürzen in a attempt to prevent the dust raised by the tracks clogging the Filzbalg filter. Three spare track links were fitted to the track guards. Krupp-built vehicles had two track links on the right front track guard and one behind the spare wheel bin on the left side. The first three triangular hangers of the hull Schürzen rails were cut to conform to the shape of the plates. Hull Schürzen of NiWerk-assembled tanks featured a distinctive cut-out on the lower edge of the first plate. This was maintained until the end of Pzkpfw IV ausf H production.

December 1943. To strengthen the hull, the 80mm-thick front plate was interlocked with the hull side extensions during production.

An order for Krupp to assemble 100 Sturmgeschütz IV assault guns per month was rescinded and the company directed to build 150 Pzkpfw IV per month. A contract with the same firm to produce the Panzerbefehlswagen IV was cancelled in January 1944 causing the company's management to later claim that this kind of indecision accounted for 300,000 lost man-hours which equated to 150 tanks.

January 1944. A Nahverteidigungswaffe, or close defence weapon, was supposed to have been mounted in the turret roof but a shortage of weapons meant that many tanks left the factories without it and it was not a standard feature until July 1944. On vehicles where the necessary aperture had been cut in the turret roof, the hole was covered by a metal plate secured by four bolts. Units in the field were ordered to apply Zimmerit to their tanks.

February 1944. It was found that large amounts of dust gathered between the Schürzen and the hull side and it was decided that the Filzbalg filters would no longer be mounted on the track guard and an improved oil filter was installed in the engine compartment. Units in the field were ordered to remove the Filzbalg filters from their tanks but many were still in place by the time of the Normandy invasion and the remains can be seen on a number of damaged tanks. Zimmerit was now applied to the hull sides as well as the other vertical surfaces. A programme of converting Pzkpfw IV ausf G tanks returned to Germany for repair to Panzerbefehlswagen command tanks began in this month.

The first Pzkpfw IV ausf J vehicles were completed and accepted by the Heereswaffenamt in February 1944 and by the following month had completely superseded the Pzkpfw IV ausf H.

..........continued on page 55

Photographed just prior to the Normandy landings, this Pzkpfw IV ausf H is from 8.Kompanie, Panzer-Regiment 3. The Fahrgestellenummer, 89272, is clearly visible on the glacis to the left of the Kugelblende machine-gun mount and this identifies this tank as a Nibelungenwerk-built vehicle. The unit insignia of 2.Panzer-Division has been applied to the other side. This Pzkpfw IV was assembled with the hull that featured a homogenous glacis and front plate of 80mm thickness and with the hull extensions interlocked with the front plate. The Filzbalg oil filter is intact and in its place on the hull side and although these were supposed to have been removed from vehicles in February 1944, many can be seen after that date.

..........continued from page 53

Assembly of the Pzkpfw IV ausf J was undertaken by Nibelungenwerk and Vomag but production at the latter was diverted to the Jagdpanzer IV in May 1944 after less than 200 tanks had been completed. The major external difference between the new version and earlier models was the deletion of the electrical turret drive (1) meaning that the exhaust muffler for the auxiliary generator, which was located on the left side of the engine compartment, was no longer fitted to the hull rear plate. Almost 2,000 Pzkpfw IV ausf J tanks were built and accepted by the end of August 1944 but as allocation records do not differentiate between production versions of particular tanks it is impossible to know how many were shipped to units in the West but it cannot have been more than 590, the total number of Pzkpfw IV tanks despatched in that period. Photographic evidence would suggest that it was considerably less (2). Major modifications incorporated into the production of the Pzkpfw IV ausf H up to July 1944 are listed below.

February 1944. The three spare track links on the hull rear plate, which had been an identifying feature of NiWerk-manufactured tanks, were no longer fitted.

March 1944. The cut-out in the first plate of the hull Schürzen was dropped.

May 1944. An earlier decision to delete the vision ports and machine-pistol ports of the turret side access hatches was implemented but the change was slow and tanks were assembled until the end of the war with these features. The raised housing over the radiator filler caps on the rear deck were simplified by having their sides squared-off.

June 1944. S-shaped hooks for the towing cables were to replace the C-shaped versions which had been a feature of Pzkpfw IV production from this month but there is strong photographic evidence that this modification was never implemented. Three sockets, or Pilze, were welded to the turret roof to hold a 2-ton jib boom that was capable of lifting large parts, such as the tank's engine (3).

July 1944. The thickness of the superstructure roof was increased to 16mm by the addition of armour plates but a plan to replace the hatches of the driver and radio operator was postponed until old stocks were exhausted. A large armoured cover was fitted to the fume extractor on the turret roof. An earlier proposal to mount a 200-litre fuel tank in place of the auxiliary generator was begun in this month when the first was fitted to Fgst Nr. 91501. Within weeks serious leaks were discovered in the tanks and installation was suspended until September.

Notes

1. Interestingly, trials conducted on captured tanks by the British showed that the manual turret traverse was as effective, and in some situations better, than the electrical drive.
2. From April 1944 a number of Pzkpfw IV ausf J tanks were converted to Artillerie-Beobachtungswagen observation tanks but as none of these had been delivered before the conclusion of the Normandy campaign they are not examined here.
3. Units in the field were ordered to fit the Pilze sockets to the turrets of their tanks in the same month.

This Nibelungenwerk-built tank is almost certainly a Pzkpfw IV ausf H, identified by the driver's hatch (A) but the Fahrgestelle Nummer which is stamped on the trackguard brace (B), and would positively define the type, is unfortunately illegible. The muzzle brake (C) is the first type fitted to the 7.5cm L/48 gun. Other features to note are: D. The interlocked homogenous front plate and hull extensions. E. The distinctive NiWerk turret Schürzen support. F. The hull Schürzen rail with its trimmed triangular hanger. G. The rectangular splash guard of the two forward hatches. H. The Bosch headlight. I. Fire extinguisher. J. C-Shaped towing hooks.

Manufactured at the Nibelungenwerk plant in September 1943, this Pzkpfw IV ausf H was one of the first tanks to be coated with Zimmerit. The texture produced by the application is made up of a pattern of small raised squares and can be seen on the turret and hull front, gun mantlet and Schürzen.

Further examples of Zimmerit application: A. Zimmerit applied with both vertical and horizontal strokes. Note how the pattern radiates from the gunner's viewing port. B. The paste applied with horizontal strokes only. In the original photograph it can be seen that this extends to the hull Schürzen. The application of Zimmerit to the hull and turret Schürzen was quickly dropped and may have only affected tanks which left the assembly lines in September and early October 1943. C. Zimmerit applied with vertical strokes to the hull of Pzkpfw IV ausf H Fgst Nr.86984 assembled by Nibelungenwerk in December 1943. D. With the production of the Pzkpfw IV ausf J the anti-magnetic mine paste seems to have been applied more sparingly, giving a thinner coat, and a different method of application, or at least a different tool, was in use. E. Another similar example of the style of Zimmerit application shown at D. This tank, Pzkpfw IV ausf J Fgst Nr.90200, was in fact completed in August or September 1944, just when Zimmerit was about to be discontinued. F. A similar style of Zimmerit application can be seen here. A number of early Pzkpfw IV ausf J tanks photographed in France, notably some from SS-Panzer-Regiment 1 and SS-Panzer-Regiment 2, had no hull Schürzen or the brackets to hold them which seem to have been purposely removed. In addition, Zimmerit had been applied to the hull sides. Note that here the paste has been also been applied to the lower hull behind the wheels.

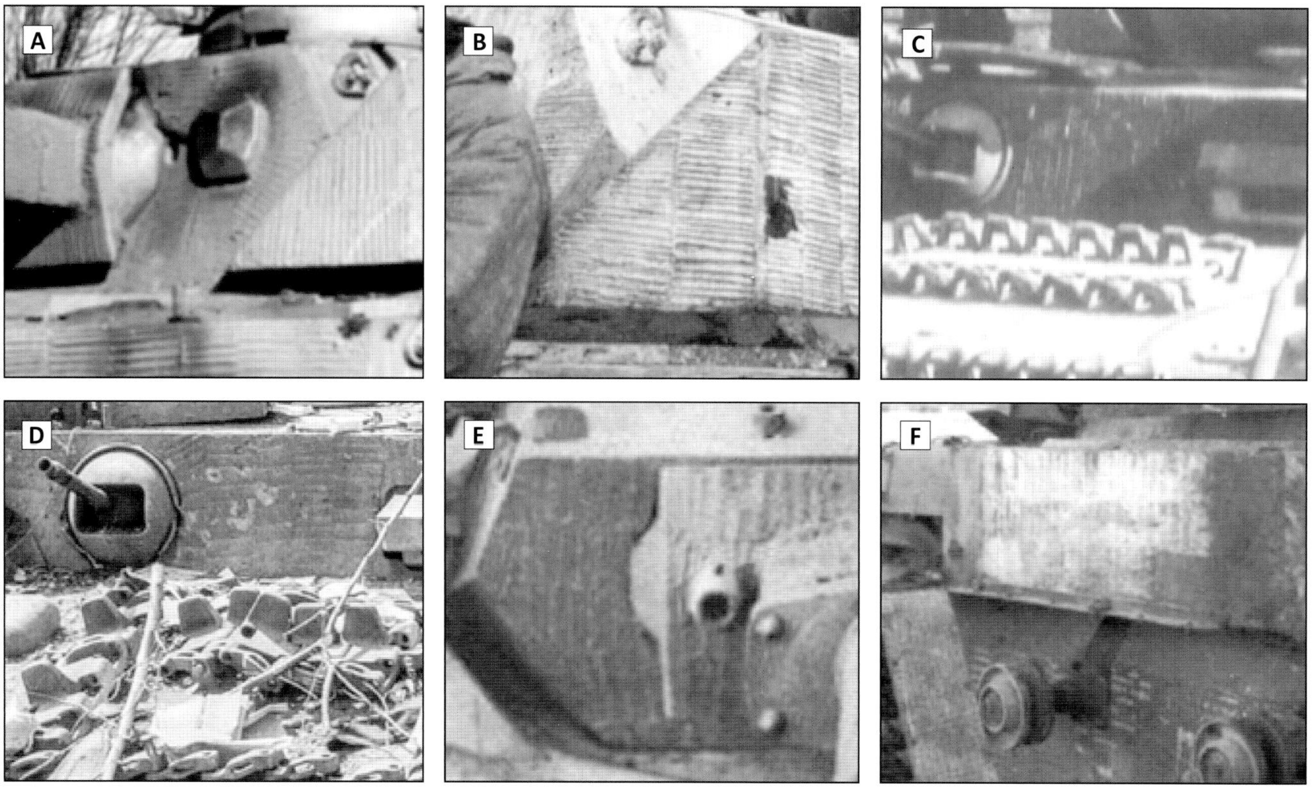

PRODUCTION AND ALLOCATION, SEPTEMBER 1943-AUGUST 1944

1. PRODUCTION

Month	Manufacturer			
	VOMAG	Krupp	Nibelungenwerk	Remarks
September 1943	75 (Pzkpfw IV ausf H)	64 (Pzkpfw IV ausf H)	150 (Pzkpfw IV ausf H)	
October 1943	85 (Pzkpfw IV ausf H)	73 (Pzkpfw IV ausf H)	170 (Pzkpfw IV ausf H)	
November 1943	69 (Pzkpfw IV ausf H)	50 (Pzkpfw IV ausf H)	120 (Pzkpfw IV ausf H)	
December 1943	90 (Pzkpfw IV ausf H)	22 (Pzkpfw IV ausf H)	210 (Pzkpfw IV ausf H)	
January 1944	99 (Pzkpfw IV ausf H)	30 (Pzkpfw IV ausf H)	170 (Pzkpfw IV ausf H)	
February 1944	54 (Pzkpfw IV ausf J)		100 (Pzkpfw IV ausf H)	19 in repair
			100 (Pzkpfw IV ausf J)	
March 1944	60 (Pzkpfw IV ausf J)		250 (Pzkpfw IV ausf J)	18 in repair
April 1944	40 (Pzkpfw IV ausf J)		259 (Pzkpfw IV ausf J)	21 in repair
				10 converted to Artillerie-Beobachtungswagen IV
May 1944	28 (Pzkpfw IV ausf J)		274 (Pzkpfw IV ausf J)	28 in repair
				22 converted to Artillerie-Beobachtungswagen IV
June 1944			300 (Pzkpfw IV ausf J)	30 in repair
				13 converted to Artillerie-Beobachtungswagen IV
July 1944			300 (Pzkpfw IV ausf J)	26 in repair
				13 converted to Artillerie-Beobachtungswagen IV
August 1944			300 (Pzkpfw IV ausf J)	19 in repair
				25 converted to Artillerie-Beobachtungswagen IV

2.ALLOCATION

Month	Unit	Quantity	Allocated by HZA	Shipped from HZA	Remarks
September 1943	21.Panzer-Division	13	6.9.1943	12.9.1943	
	21.Panzer-Division	17	6.9.1943	14.9.1943	
	Inspekteur der Panzertruppen	10	28.9.1943	24.9.1943	Panzer-Lehr-Regiment
	21.Panzer-Division	23	30.9.1943		21 tanks diverted to 14.Pz-Div
	Inspekteur der Panzertruppen	15	5.10.1943	8.10.1943	1.Komp., Panzer-Lehr-Regiment.
October 1943	9.SS-Panzer-Division	8	28.10.1943	5.10.1943	
	10.SS-Panzer-Division	8	28.10.1943	5.10.1943	
	12.SS-Panzer-Division	8	28.10.1943	5.10.1943	
	1.SS-Panzer-Division	14	26.10.1943	5.10.1943	
	1.SS-Panzer-Division	15	26.10.1943	5.10.1943	
November 1943	LVIII.Reserve-Panzerkorps	10	6.11.1943	13.11.1943	Included in October list
	9.SS-Panzer-Division	19	5.11.1943	12.11.1943	
	12.SS-Panzer-Division	16	5.11.1943	11.11.1943	
	12.SS-Panzer-Division	10	5.11.1943	13.11.1943	
December 1943	9.SS-Panzer-Division	22	4.12.1943	8.12.1943	
	10.SS-Panzer-Division	19	1.12.1943	4.12.1943	
	12.SS-Panzer-Division	6	5.12.1943	3.12.1943	
	21.Panzer-Division	4	10.12.1943	31.12.1943	
	LVIII.Reserve-Panzerkorps	6	17.12.1943	22.1.1944	
	LVIII.Reserve-Panzerkorps	14	24.12.1943	6.1.1944	
	LVIII.Reserve-Panzerkorps	9	24.12.1943	10.1.1944	
	LVIII.Reserve-Panzerkorps	8	24.12.1943	15.1.1944	
	Panzer-Lehr-Regiment	9	31.12.1943	6.1.1944	
	10.SS-Panzer-Division	22	30.12.1943	7.1.1944	
	12.SS-Panzer-Division	20	30.12.1943	5.1.1944	
January 1944	12.SS-Panzer-Division	15	11.1.1944	14.1.1944	
	12.SS-Panzer-Division	15	11.1.1944	20.1.1944	93 shipped in total
	2.Panzer-Division	17	12.1.1944	20.1.1944	
	2.Panzer-Division	17	12.1.1944	20.1.1944	
	2.Panzer-Division	11	12.1.1944	20.1.1944	
	21.Panzer-Division	9	22.1.1944	26.1.1944	
	21.Panzer-Division	12	22.1.1944	29.1.1944	78 shipped in total
	Panzer-Lehr-Division	15	29.1.1944	2.2.1944	
	Panzer-Lehr-Division	15	29.1.1944	3.2.1944	
February 1944	2.Panzer-Division	18	7.2.1944	13.2.1944	Three separate shipments made to the
	2.Panzer-Division	17	7.2.1944	16.2.1944	division on the same day with 98 tanks
	2.Panzer-Division	18	7.2.1944	17.2.1944	shipped in total up to that time
	Panzer-Lehr-Div	38	8.2.1944	26.2.1944	
	Panzer-Lehr-Div	30	18.2.1944	27.2.1944	98 shipped in total
	21.Panzer-Division	20	18.2.1944	5.3.1944	
March 1944	2.SS-Panzer-Division	25	27.3.1944	27.3.1944	
	2.SS-Panzer-Division	25	28.3.1944	28.3.1944	
	179.Reserve-Panzer-Division	17	No date given	31.3.1944	Possibly never despatched
	179.Reserve-Panzer-Division	10	No date given	31.3.1944	Possibly never despatched
	9.SS-Panzer-Division	5	30.3.1944	16.4.1944	
	10.SS-Panzer-Division	5			Cancelled
April 1944	179.Reserve-Panzer-Division	15	1.4.1944	3.4.1944	Included in March list
	179.Reserve-Panzer-Division	15	1.4.1944	3.4.1944	Included in March list
	155.Reserve-Panzer-Division	20	1.4.1944	8.4.1944	Included in March list
	155.Reserve-Panzer-Division	20	1.4.1944	1.4.1944	Included in March list
	155.Reserve-Panzer-Division	18	1.4.1944	14.4.1944	Probably to 9.Panzer-Division
	273.Reserve-Panzer-Division	18	1.4.1944	14.4.1944	Probably to 11.Panzer-Division
	273.Reserve-Panzer-Division	18	1.4.1944	14.4.1944	Probably to 11.Panzer-Division
	273.Reserve-Panzer-Division	18	1.4.1944	15.4.1944	Probably to 11.Panzer-Division
	21.Panzer-Division	14			Cancelled, shipment
					diverted to 1.SS-Pz-Div
	116.Panzer-Division	10			Cancelled, shipment
					diverted to 1.SS-Pz-Div
May 1944	1.SS-Panzer-Division	14	2.5.1944	7.5.1944	Included in April lists
	1.SS-Panzer-Division	10	2.5.1944	7.5.1944	Included in April lists
	1.SS-Panzer-Division	11	6.5.1944	7.5.1944	
	21.Panzer-Division	14	15.5.1944	24.5.1944	
	116.Panzer-Division	10	15.5.1944	29.5.1944	
June 1944	1.SS-Panzer-Division	33	2.6.1944	8.6.1944	
	1.SS-Panzer-Division	20	8.6.1944	17.6.1944	
	2.SS-Panzer-Division	24	8.6.1944	16.6.1944	
	Nachschub West	17	21.6.1944	8.7.1944	21.Panzer-Division
		11	21.6.1944	8.7.1944	Panzer-Lehr-Division
		17	21.6.1944	8.7.1944	12.SS-Panzer-Division
July 1944	Nachschub West	12	30.7.1944	10.8.1944	Panzer-Lehr-Division
		10	30.7.1944	10.8.1944	21.Panzer-Division
		12	30.7.1944	10.8.1944	12.SS-Panzer-Division
August 1944	Nachschub West	6	21.8.1944	21.8.1944	15.Panzergrenadier-Division
		8	19.8.1944	19.8.1944	
		17	1.9.1944	1.9.1944	1.SS-Panzer-Division

PANZER-REGIMENT, NOVEMBER 1943

All formations of the German Army were organised according to detailed tables of establishment referred to as Kriegstärkenachweisung (KstN) which showed the official composition of a unit, listing the exact number of personnel and type of equipment from vehicles to small arms. The tables were numbered and dated, the latter being all-important as some numbers were duplicated. The KstN were accompanied by Kriegsausrüstungsnachweisung (KAN) which listed minor items in detail such as tools and fuel containers but these need not concern us here. In November 1943, in an effort to standardise the organisation of the Panzer regiments and facilitate the introduction of the Panther battalions, new KstN were authorised for units in the process of formation or rebuilding. Existing formations were usually instructed to implement new establishments as circumstances and equipment allowed. Although amended KstN were issued in April 1944 a number of regiments which fought in the Normandy battles, for example Panzer-Regiment 3, did so with battalions built following the November 1943 organisation. It should be remembered that these tables of establishment represent the ideal and that few of the regiments engaged in the battles of summer 1944 were at their authorised strength.

Stab und Stabs-Kompanie eines Panzer-Regiments KstN 1103 von 1.11.1943

- Regiments-Stab
- Stabs-Kompanie
  - Nachrichten Zug: Befehlspanzer V, Befehlspanzer V, Befehlspanzer V
  - Aufklärungs Zug: Pzkpfw IV, Pzkpfw IV, Pzkpfw IV, Pzkpfw IV

II.Abteilung
Stabs-Kompanie einer Panzer-Abteilung
KstN 1150b von 1.11.1943

- Nachrichten-Zug: Befehlspanzer III, Befehlspanzer III, Befehlspanzer III
- Aufklärungs-Zug: Pzkpfw IV, Pzkpfw IV, Pzkpfw IV, Pzkpfw IV, Pzpfkw IV
- Flak-Zug

The only units to report any Pzkpfw III tanks in June 1944 were Panzer-Regiment 22, Panzer-Abteilung (Fkl) 301 and Panzer Ausbildungs und Ersatz Abt. 100. In July, Panzer-Regiment 16 had thirteen tanks armed with a variety of guns.

Each battalion was authorise an anti-aircraft platoon with six Sdkfz 10/4 halftracks armed with either the 2.0cm Flak 30 or Flak 38 gun and three halftracks with the guns removed that acted as ammunition carriers

5.Kompanie
Mittlere Panzer-Kompanie a
KstN 1175a von 1.11.1943

- Gruppe Führer: Pzkpfw IV, Pzkpfw IV
- 1.Zug: Pzkpfw IV, Pzkpfw IV, Pzkpfw IV, Pzkpfw IV, Pzkpfw IV
- 2.Zug: Pzkpfw IV, Pzkpfw IV, Pzkpfw IV, Pzkpfw IV, Pzkpfw IV
- 3.Zug: Pzkpfw IV, Pzkpfw IV, Pzkpfw IV, Pzkpfw IV, Pzkpfw IV
- 4.Zug: Pzkpfw IV, Pzkpfw IV, Pzkvw IV, Pzkpfw IV, Pzkpfw IV

6.Kompanie
Mittlere Panzer-Kompanie a
KstN 1175a von 1.11.1943
As for 5.Kompanie

7.Kompanie
Mittlere Panzer-Kompanie a
KstN 1175a von 1.11.1943
As for 5.Kompanie

8.Kompanie
Mittlere Panzer-Kompanie a
KstN 1175a von 1.11.1943
As for 5.Kompanie

Panzer-Werkstatt-Zug
KstN 1185 von 1.6.1942

I.Abteilung
Stabs-Kompanie einer Panzer-Abteilung Panther
KstN 1150a von 1.11.1943

- Nachrichten-Zug
- Aufklärungs-Zug
- Flak-Zug

1.Kompanie
Mittlere Panzer-Kompanie Panther
KstN 1177 von 1.11.1943

2.Kompanie
Mittlere Panzer-Kompanie Panther
KstN 1177 von 1.11.1943

3.Kompanie
Mittlere Panzer-Kompanie Panther
KstN 1177 von 1.11.1943

4.Kompanie
Mittlere Panzer-Kompanie Panther
KstN 1177 von 1.11.1943

Panzer-Werkstatt-Zug Panther
KstN 1185a von 10.1.1943

The companies of I.Abteilung were organised as for il.Abteilung except that all the tanks, including the command Panzers, were replaced by Pzkpfw V Panthers

Some regiments had their Pzkpfw IV tanks concentrated in I.Abteilung, contrary to the official orders, and these are mentioned on the unit histories. Many German military terms are difficult to translate into English and they are presented here in the form in which they appeared in the original documents.

PANZER-LEHR-REGIMENT 130, MARCH 1944

It was planned that this regiment, and its parent division, would be the first of the Panzer units to be organised under the frei Gliederung (fG) concept and special KstN were issued as shown below. At the end of March 1944, II.Abteilung was at full strength but, as mentioned in the unit history, a number of tanks were handed over to Panzer-Regiment 22. On 1 June 1944 the regiment reported that 5.Kompanie and 6.Kompanie each contained twenty-one Pzkpfw IV tanks while 7.Kompanie and 8.Kompanie had their full allocation. Twelve Flakpanzer 38t were sent to the regiment on 12 December 1943 but this allocation seems to have been cancelled as twelve more were shipped on 10 February 1944. I would assume that the KstN was the same as that authorised for Panzer-Regiment 22 at about the same time but I cannot confirm if these anti-aircraft tanks were on hand in June. In a status report dated 1 February 1944 the division's commander stated that ten Pzkpfw V Panthers were on hand and in serviceable condition although just three were authorised.

Stab und Stabs-Kompanie eines Panzer-Regiments KstN 1103 von 1.11.1943

- Regiments-Stab
- Stabs-Kompanie
  - Nachrichten Zug: Befehlspanzer V, Befehlspanzer V, Befehlspanzer V
  - Aufklärungs Zug: Pzkpfw IV, Pzkpfw IV, Pzkpfw IV, Pzkpfw IV

II.Abteilung
Stab Panzer-Abteilung IV
KstN fG von 1.11.1943

Note that the KstN shown here are not numbered as was sometimes the case with temporary or trial establishments.

I.Abteilung
Stab Panzer-Abteilung Panther
KstN fG von 1.11.1943

- Stabs-Kompanie Panzer-Abteilung Panther KstN fG von 1.11.1943

The organisation for the regiment's I.Abteilung authorised a Nachrichten-Zug and Aufklärungs-Zug with eight Panther tanks, three companies with twenty-two Panthers each and one company equipped with twenty-two Jagdpanther tank destroyers. In addition, the Tigers of Panzer-Kompanie (Funklenk) 316 were to be attached to the regimental headquarters. This establishment was never realised and the division fought the Normandy battles with the Panthers of Panzer-Regiment 6 in place of its I.Abteilung.

II.Abteilung:

- Stabs-Kompanie Panzer-Abteilung IV KstN fG von 1.11.1943
  - Nachrichten-Zug (three tanks)
    Possibly Panzerbefehlswagen IV although there is no direct reference in the Gliederung or strength returns.
  - Aufklärungs-Zug: Pzkpfw IV, Pzkpfw IV, Pzkpfw IV, Pzkpfw IV, Pzkpfw IV
- 5.Kompanie Panzer-Kompanie IV KstN fG von 1.11.1943
  - Gruppe Führer: Pzkpfw IV, Pzkpfw IV
  - 1.Zug: Pzkpfw IV, Pzkpfw IV, Pzkpfw IV, Pzkpfw IV, Pzkpfw IV
  - 2.Zug: Pzkpfw IV, Pzkpfw IV, Pzkpfw IV, Pzkpfw IV, Pzkpfw IV
  - 3.Zug: Pzkpfw IV, Pzkpfw IV, Pzkpfw IV, Pzkpfw IV, Pzkpfw IV
  - 4.Zug: Pzkpfw IV, Pzkpfw IV, Pzkpfw IV, Pzkpfw IV, Pzkpfw IV
- 6.Kompanie Panzer-Kompanie IV KstN fG von 1.11.1943 — *As for 5.Kompanie*
- 7.Kompanie Panzer-Kompanie IV KstN fG von 1.11.1943 — *As for 5.Kompanie*
- 8.Kompanie Panzer-Kompanie IV KstN fG von 1.11.1943 — *As for 5.Kompanie*
- Versorgungs-Kompanie frei Panzer-Abteilung KstN fG von 1.11.1943

PANZER-REGIMENT 22, MAY 1944

On 20 May 1944 Panzer-Regiment 100 of 21.Panzer-Division was ordered to be reorganised according to the frei Gliederung system and renamed Panzer-Regiment 22. The regiment's I.Abteilung was to be made up of four companies, each with seventeen Pzkpfw IV tanks. At that time I.Abteilung still had a number of French tanks on hand but these were gradually relinquished and it was planned that each of the four companies would receive fourteen Pzkpfw IV vehicles. By the time of the Normandy landings the French tanks had gone but the full authorisation for II.Abteilung was never reached and the actual situation is shown here. Note that I have listed the KstN as they were given in the orders dated 11 October 1943 issued shortly after the first Pzkpfw IV tanks arrived.

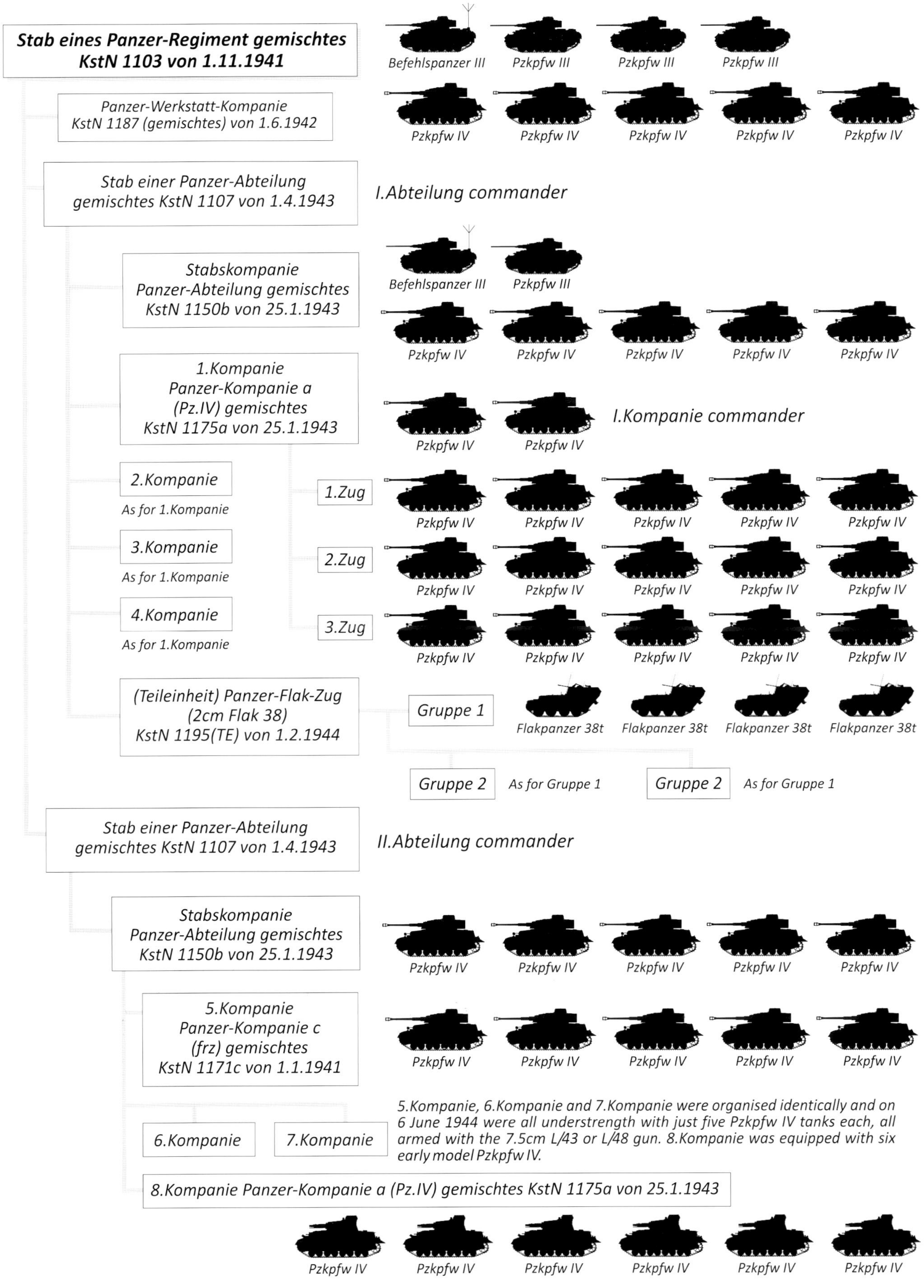

PANZER-REGIMENT, APRIL 1944

Efforts to simplify the organisation of Panzer units began as early as December 1943, with the formation of the Panzer-Lehr-Division, and by April 1944 instructions had been issued to institute the so-called frei Gliederung, or free organisation, concept where support and transport elements were taken from companies and concentrated at regimental level. New KstN were introduced which were to take effect from 1 April 1944 for units under the control of the Inspekteur der Panzertruppen and establishments for other units were disseminated in the following months. Waffen-SS regiments were ordered to adopt the changes in May 1944. The final instruction to implement the frei Gliederung organisations was not issued until 3 August and did not include Panzer-Regiment 22 and Panzer-Lehr-Regiment 130.

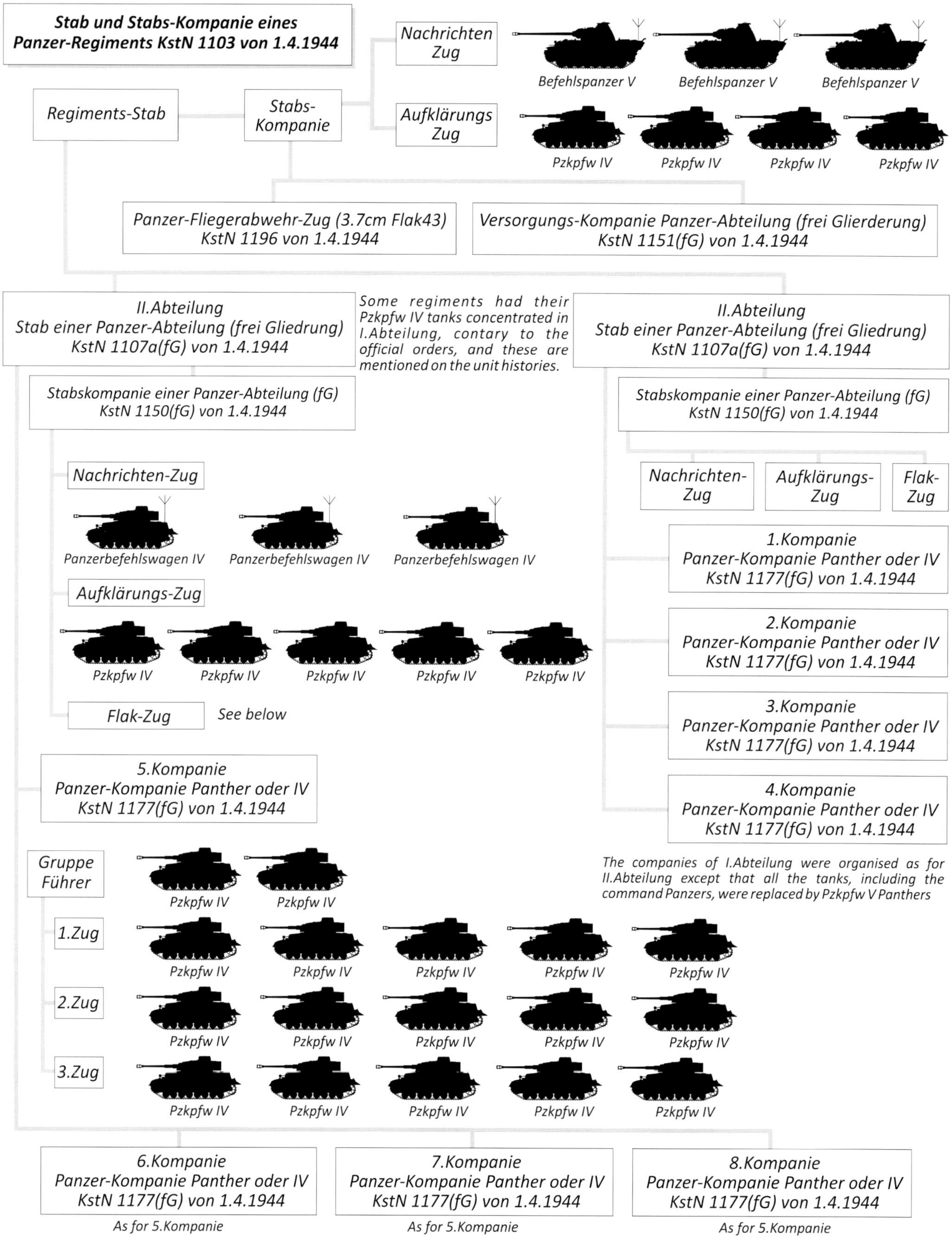

The provisional organisation introduced from 1 April 44 retained an anti-aircraft platoon at battalion level made up of three Sdkfz 7/1 halftracks armed with the 2.0cm Flak 38 quad mounting. An additional platoon was added for the regiment comprising four groups each equipped with two Sdkfz 161/3 Möbelwagen. But none saw action in the West until August 1944 when schwere SS-Panzer-Abteilung 101 reported four on hand and the Flakpanzer 38t was far more common.

II.ABTEILUNG, SS-PANZER-REGIMENT 12, JUNE 1944

As mentioned in the text this regiment's II.Abteilung contained a fifth company, referred to as 9.Kompanie, and this was never reported officially to either the SS-Führungshauptamt, the SS Main Office, or the Inspekteur der Panzertruppen. But there is ample photographic evidence for the existence of 9.Kompanie and it is mentioned in several authoritative sources including the battalion's Kriegstagebuch and the post-war history written by Hubert Meyer. In addition to its unique organisation the battalion's tanks were identified by a non-standard numbering system which is shown below and has been pieced together from mentions in official records and surviving photographs. The presence of the tanks shown as light grey silhouettes is unconfirmed but assumed on the basis that a higher number was used. It is also possible that the regimental headquarters had a number of Pzkpfw IV tanks on hand and these may have been numbered as RA1, RA2 etc. The operational life of 9.Kompanie was short and due to battle casualties it was absorbed by 5.Kompanie on 1 August 1944 and the battalion reverted to a four-company establishment.

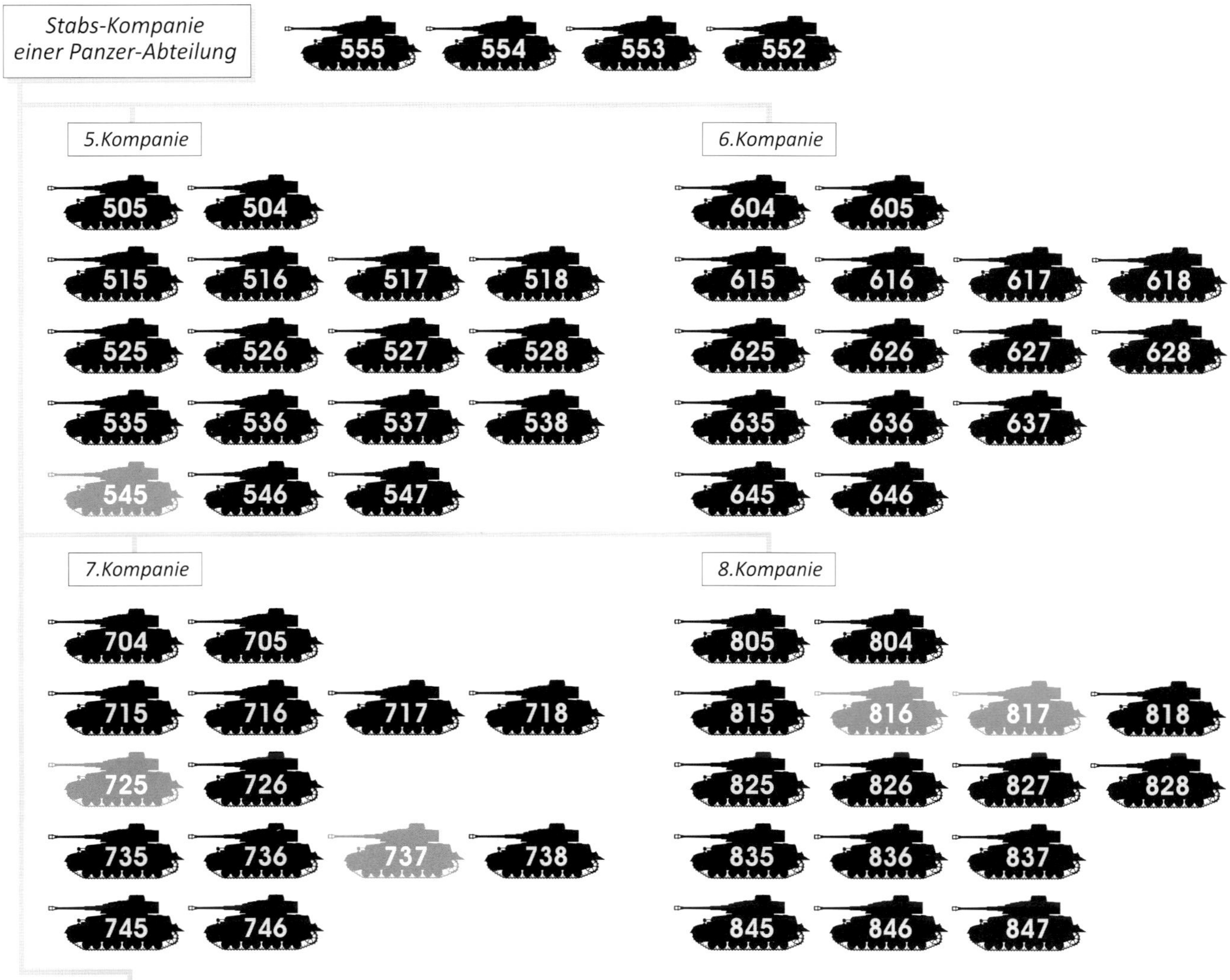

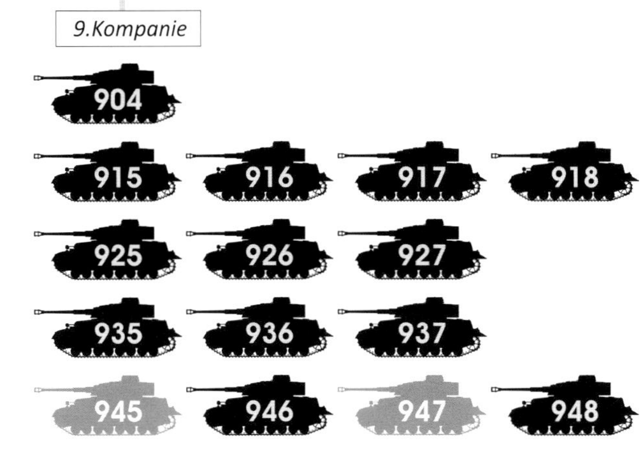

A directive issued by the regimental headquarters dated 4 January 1944 stipulated that the numbers to be used for the Abteilung-Stab were 555 and 554 for the commander and his adjutant, 553 for the signals officer and 556 to 560 for the Aufklärungs-Zug, or reconnaissance platoon. The tanks of the four companies were to be identified by the company number followed by 05 and 04 for the Kompanie-Führer and Truppe-Führer and each company was to contain four platoons. Each platoon was to contain five tanks and these were to be marked with the company number followed by 15 to 19 for the first platoon, 25 to 29 for the second and so on. The decision to create 9.Kompanie must have been made at some time after this order and the number of tanks in each platoon reduced to four. The complete absence of any photographic evidence to support a number ending in 9 would seem to confirm this. The photograph above shows a Pzkpfw IV disabled and abandoned in the Falaise Pocket in August 1944 which may have some connection to the regimental headquarters of SS-Panzer-Regiment 12 as mentioned earlier.

Dragon Models Ltd
B1-10/F., 603-609 Castle Peak Rd,
Kong Nam Industrial Building,
Tsuen Wan, N. T., Hong Kong
www.dragon-models.com

Tamiya Inc
Shizuoka City, Japan
www.tamiya.com

Trumpeter/Hobby Boss
NanLong Industrial Park, SanXiang,
Zhong Shan, GuangDong, China
www.trumpeter-china.com
www.hobbyboss.com

Academy Plastic Models
521-1, Yonghyeon-dong, Uijeongbu-si,
Gyeonggi-do, Korea
www.academy.co.kr

Hobby Fan/ AFV Club
6F., No.183, Sec. 1, Datong Rd, Xizhi City,
Taipei County 221, Taiwan
www.hobbyfan.com

Zvezda (Zvezda-America)
www.zvezda-usa.com
Note that the Russian catalogue is not the same as the US version.

Italeri S.p.A.
via Pradazzo 6/b,
40012 Calderara di Reno, Bologna, Italy
www.italeri.com

Takom
www.takom-world.com

Border Model
Xizhang Industrial Park, Huishan District,
Wuxi City, P.R. China
www.bordermodel.com

Airfix
www.airfix.com

Rye Field Models
www.ryefield-model.com
An almost non-existent website. I would recommend one of the on-line retailers.

Hauler
Jan Sobotka,
Moravská 38, 620 00 Brno,
Czech Republic
www.hauler.cz

Voyager
Room 501, No.411 4th Village,
SPC Jinshan District, Shanghai 200540
P.R.China
www.voyagermodel.com

Griffon Model
Suite 501, Bldg 01, 418 Middle Longpan Rd,
Nanjing, P.R. China
www.griffonmodel.com

Aber
ul. Jalowcowa 15, 40-750 Katowice, Poland
www.aber.net.pl

E.T. Model
www.etmodeller.com

Friulmodel
H 8142. Urhida, Nefelejcs u. 2., Hungary
www.friulmodel.hu

Modelkasten
Chiyoda-ku Kanda, Nishiki-Cho 1-7, Tokyo,
Japan
www.modelkasten.com
Very difficult to navigate but worthwhile.

Theodoros Kalamatas Modelling Worshop
www.facebook.com/Theodoros-Kalamatas-Modelling-Workshop

Eduard Model Accessories
Mirova 170, 435 21 Obrnice,
Czech Republic
www.eduard.com

Master Club
www.masterclub.ru
It appears that this firm is closely associated with Armour35, a Russian mail-order firm.

Model Artisan Mori
Yasutsugu Mori,
Maison Suiryu 302, Kunoshiro-cho 1-10,
Yokkaichi-City, Mie 510-0072, Japan
www.artisanmori.web.fc2.com

Model Factory Hiro
Yubinbango 121-0063 Adachi-ku, Tokyo,
Higashihokima 2-Chome, 3-8 Japan
www.modelfactoryhiro.com

RB Model
Powstancow Wlkp.29B,
64-360 Zbaszyn,
Poland
www.rbmodel.com

M Workshop Singapore
91 Bencoolen St, Sunshine Plaza 01-58,
Singapore
www.themworkshop.com

ROCHM Model
www.rochmmodel.com
rochmmodel@gmail.com
This company has a huge selection of parts and accessories for 1/35 scale Panther models.

Royal Model
Via E. Montale, 19-95030 Pedara, Italy
www.royalmodel.com

In writing this book I referred extensively to the *Panzertracts* series of booklets created by the late Thomas Jentz and Hilary Doyle and to a lesser extent, the Pzkpfw IV studies of Waldemar Trojca. Another invaluable resource is Thomas Jentz's *Panzertruppen* books which I would recommend to any reader with an interest in German armour. Although dated, *Panzer IV und seine Abarten* by Walter Spielberger, which has also been published in English, was still helpful. The information on unit allocations was all taken from original Zufürung, or supply lists, compiled by the Heereszeugamt between May 1943 and April 1945. I also relied heavily on the work of Martin Block and the late Ron Klages whose research on vehicle allocations, including those which actually arrived at their destination, would fill many volumes. I would like to thank the modellers who generously allowed me to publish the images of their work as well as Lisa Hooson and Stephen Chumbley, my editors at Pen & Sword. As always, I am indebted to Karl Berne, Valeri Polokov and J.Howard Parker for their invaluable assistance with the photographs and period insignia.

A Nibelungenwerk-built Pzkpfw IV ausf H of 6.Kompanie, SS-Panzer-Regiment 12 photographed in the spring of 1944 when the regiment was training in Belgium. Note the name 'Gerti' painted on either side of the commander's cupola.